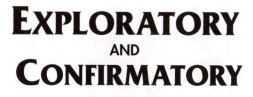

EXPLORATORY
AND
CONFIRMATORY
FACTOR ANALYSIS

EXPLORATORY
AND
CONFIRMATORY
FACTOR ANALYSIS

Understanding Concepts and Applications

BRUCE THOMPSON

American Psychological Association • Washington, DC

Published by
American Psychological Association
750 First Street, NE
Washington, DC 20002
www.apa.org

To order
APA Order Department
P.O. Box 92984
Washington, DC 20090-2984
Tel: (800) 374-2721
Direct: (202) 336-5510
Fax: (202) 336-5502
TDD/TTY: (202) 336-6123
Online: www.apa.org/books/
E-mail: order@apa.org

In the U.K., Europe, Africa, and the Middle East, copies may be ordered from
American Psychological Association
3 Henrietta Street
Covent Garden, London
WC2E 8LU England

Typeset in Goudy by World Composition Services, Inc., Sterling, VA

Printer: Data Reproductions, Auburn Hills, MI
Cover Designer: Veronica Breza, Sterling, VA
Technical/Production Editor: Dan Brachtesende

The opinions and statements published are the responsibility of the author, and such opinions and statements do not necessarily represent the policies of the American Psychological Association.

Library of Congress Cataloging-in-Publication Data

Thompson, Bruce, 1951–
Exploratory and confirmatory factor analysis : understanding concepts and applications / Bruce Thompson.—1st ed.
p. cm.
Includes bibliographical references.
ISBN 1-59147-093-5
1. Factor analysis—Textbooks.   I. Title.

BF39.2.F32T48 2004
150'.1'5195354—dc22

2003022596

British Library Cataloguing-in-Publication Data
A CIP record is available from the British Library.

Printed in the United States of America
First Edition

CONTENTS

PREFACE

Investigation of the structure underlying variables (or people, or time) has intrigued social scientists since the early origins of psychology. Conducting one's first factor analysis can yield a sense of awe regarding the power of these methods to inform judgment regarding the dimensions underlying constructs.

This book presents the important concepts required for implementing two disciplines of factor analysis: exploratory factor analysis (EFA) and confirmatory factor analysis (CFA). The book may be unique in its effort to present both analyses within the single rubric of the general linear model. Throughout the book canons of best factor analytic practice are presented and explained.

The book has been written to strike a happy medium between accuracy and completeness versus overwhelming technical complexity. An actual data set, randomly drawn from a large-scale international study (cf. Cook & Thompson, 2001) involving faculty and graduate student perceptions of academic libraries, is presented in Appendix A. Throughout the book different combinations of these variables and participants are used to illustrate EFA and CFA applications.

Real understanding of statistical concepts comes with applying these methods in one's own research. The next-best learning opportunity involves analyzing real data provided by others. You are encouraged to replicate the analyses reported in the book and to conduct methodological variations on the reported analyses. You also may find it useful to conduct complementary analyses to those reported here (e.g., if 11 variables from 100 faculty are analyzed, conduct the same analysis using data from the 100 graduate students).

There are also other data sets that have a long history of use in heuristics in the factor analytic context. For example, the data reported by Holzinger and Swineford (1939) are widely reported. Readers are encouraged to access and analyze these and other data sets.

To improve the instructional effectiveness of the book, some liberties have been taken in using American Psychological Association (APA) style. For example, to distinguish between *score world* and *area world* statistics, area world statistics are often presented with percentage signs, even across the rows of a single table, for emphasis and clarity. And more than the usual number of decimal places may be reported, to help the reader compare reports in the book with the related reports in computer outputs, or to distinguish one block of results from another by using different numbers of decimal places for different results.

I appreciate in this regard the latitude afforded me by the APA production staff. I also appreciate the opportunity to write the book in my own voice. This book has been written as if I were speaking directly to you, the student, in my own office or classroom. Fortunately, I have a very big office!

I also appreciate the thoughtful suggestions of Randy Arnau, University of Southern Mississippi, Susan Cromwell, Texas A&M University, and Xitao Fan, University of Virginia, on a draft of the book. Notwithstanding their suggestions, of course I remain responsible for the book in its final form.

EXPLORATORY
AND
CONFIRMATORY
FACTOR ANALYSIS

1

INTRODUCTION TO
FACTOR ANALYSIS

The early 1900s were exciting times for social scientists. Psychology was emerging as a formal discipline. In France, Alfred Binet created a measure of intellectual performance that was the forerunner to modern IQ tests. But from the very outset there were heated controversies regarding definitions of intelligence and related questions about how to measure intelligence.

Some psychologists took the view that intelligence was a single general ability. Those taking this view presumed that individuals who were very bright in one area of performance tended to perform well across the full range of intellectual tasks. This came to be called the general, or "G," theory of intelligence. Other psychologists argued that individuals may be proficient at one task, but that expertise in one area has little or no predictive ability as regards other areas of cognitive performance. These latter psychologists argued that intelligence involved specific factors that correlated minimally with each other.

However, in these early years psychologists who wished to test their divergent views had few analytic tools with which to do so. To address this void, Spearman (1904) conceptualized methods that today we term factor analysis. In the intervening decades, numerous statistical developments mean that today the analyst confronts an elaborate array of choices when using these methods.

Of course, for many decades factor analysis was infrequently used. In the era of statistical analysis via hand-cranked calculators, the computations of factor analysis were usually overwhelming. In part this was also because responsible researchers conducting analyses prior to the advent of modern computers felt obligated to calculate their results repeatedly, until at least two analyses of the data yielded the same results!

Happily, today the very, very complicated calculations of modern factor analysis can be accomplished on a microcomputer in only moments, and in a user-friendly point-and-click environment. As a consequence, today factor analytic reports comprise as many as 18% to 27% of the articles in some journals (Fabrigar, Wegener, MacCallum, & Strahan, 1999; Russell, 2002).

PURPOSES OF FACTOR ANALYSIS

Factor analytic methods can be used for various purposes, three of which are noted here. First, when a measure and its associated scoring keys have been developed, *factor analysis can be used to inform evaluations of score validity*.

Validity questions focus on whether scores measure "the correct something." Measures producing scores that are completely random measure "nothing" (Thompson, 2003). Such tools yield scores that are perfectly unreliable. We never develop tests to measure nothing, because we have readily available tools, such as dice and roulette wheels, that we can use to obtain random scores. Obviously, for scores intended to measure something (e.g., intelligence, self-concept), the scores must be reliable. But reliability is only a necessary—not a sufficient—condition for validity.

Validity questions involve issues such as, "Does the tool produce scores that seem to measure the intended dimensions?" and "Are items intended only to measure a given dimension actually measuring and only measuring that dimension?" These are exactly the kinds of questions that factor analysis was created to confront. So it is not surprising that historically "construct validity has [even] been spoken of as . . . 'factorial validity' " (Nunnally, 1978, p. 111).

Joy Guilford's discussion some 50 years ago illustrates this application:

Validity, in my opinion is of two kinds. . . . The *factorial validity* [italics added] of a test is given by its . . . [factor weights on] meaningful, common, reference factors. This is the kind of validity that is really meant when the question is asked "Does this test measure what it is supposed to measure?" A more pertinent question should be "What does this test measure?" The answer then should be in terms of factors.

. . . I predict a time when any test author will be expected to present information regarding the factor composition of his [*sic*] tests. (Guilford, 1946, pp. 428, 437–438)

Although modern validity language does not recognize a type of validity called "factorial validity," factor analysis nevertheless is helpful in addressing construct validity questions.

Thus, Gorsuch (1983) has noted that "a prime use of factor analysis has been in the development of both the operational constructs for an area and the operational representatives for the theoretical constructs" (p. 350). And Nunnally (1978) suggested that "factor analysis is intimately involved with questions of validity. . . . Factor analysis is at the heart of the measurement of psychological constructs" (pp. 112–113).

Second, *factor analysis can be used to develop theory regarding the nature of constructs*. At least historically, empirically driven theory development was sometimes held in high regard. In some of these applications, numerous different measures are administered to various samples, and the results of factor analysis then are used to specify construct dimensions. This application, unlike the first one, is inductive.

A classic example is the structure of intellect model developed by Guilford (1967). Guilford administered dozens of tests of cognitive abilities over years to derive his theory, in which he posited that intelligence consists of more than 100 different abilities that are independent of each other.

Third, *factor analysis can be used to summarize relationships in the form of a more parsimonious set of factor scores that can then be used in subsequent analyses* (e.g., analysis of variance, regression, or descriptive discriminant analysis). In this application, unlike the first two, the factor analysis is only an intermediate step in inquiry, and not the final analysis. Using fewer variables in substantive analysis tends to conserve degrees of freedom and improve power against Type II error. Furthermore, factor scores can be computed in such a manner that the unreliable variance in the original variables tends to be discarded once the original scores are reexpressed in a smaller set of factor scores.

TWO MAJOR CLASSES OF FACTOR ANALYSIS

There are actually two discrete classes of factor analysis: exploratory factor analysis (EFA) and confirmatory factor analysis (CFA). Analyses such as those originally proposed by Spearman (1904) have now come to be called exploratory factor analysis. In EFA, the researcher may not have any specific expectations regarding the number or the nature of underlying constructs or factors. Even if the researcher has such expectations (as in a validity investigation), EFA does not require the researcher to declare these expectations, and the analysis is not influenced by these expectations.

Confirmatory factor analysis methods were developed much more recently (cf. Jöreskog, 1969). These analyses require that the researcher must have specific expectations regarding (a) the number of factors, (b) which variables reflect given factors, and (c) whether the factors are correlated. CFA explicitly and directly tests the fit of factor models.

Researchers without theories *cannot* use CFA, but researchers with theories usually find CFA more useful than EFA. Confirmatory factor analysis is more useful in the presence of theory because (a) the theory is directly tested by the analysis and (b) the degree of model fit can be quantified in various ways. Of course, some researchers begin an inquiry with specific theoretical expectations, but after invoking CFA find that their expectations were wildly wrong, and may then revert to EFA procedures.

Although factor analysis has been characterized as "one of the most powerful methods yet for reducing variable complexity to greater simplicity" (Kerlinger, 1979, p. 180) and as "the furthest logical development and reigning queen of the correlational methods" (Cattell, 1978, p. 4), the methods have been harshly criticized by others (cf. Armstrong, 1967). For example, with respect to research in the area of communication, Cronkhite and Liska (1980) noted:

> Apparently, it is so easy to find semantic scales which seem relevant to [information] sources, so easy to name or describe potential/hypothetical sources, so easy to capture college students to use the scales to rate the sources, so easy to submit those ratings to factor analysis, so much fun to name the factors when one's research assistant returns with the computer printout, and so rewarding to have a guaranteed publication with no fear of nonsignificant results that researchers, once exposed to the pleasures of the factor analytic approach, rapidly become addicted to it. (p. 102)

However, these criticisms apply more to EFA than to CFA, because the only danger in EFA would be that factors might not be interpretable, in which case the results may be regarded as negative. But CFA presumes the invocation of specific expectations, with the attendant possibility that the predetermined theory in fact does not fit, in which case the researcher may be unable to explain the relationships among the variables being analyzed. Both EFA and CFA remain useful today, and our selection between the two classes of factor analysis generally depends on whether we have specific theory regarding data structure.

But even researchers who only conduct research in the presence of such theory must understand EFA. Mastery of CFA requires understanding of EFA, which was the historical precursor to CFA. Indeed, a fundamental purpose of this book is to teach the reader what is common and what is different across these two classes of factor analysis.

Understanding EFA also may facilitate deeper insight into other statistical analyses. An important idea in statistics is the notion of the general linear model (GLM; cf. Bagozzi, Fornell, & Larcker, 1981; Cohen, 1968; Knapp, 1978). Central to the GLM are realizations that *all* statistical analyses are correlational and can yield effect sizes analogous to r^2, and apply weights to measured variables to obtain scores on the composite variables that are actually the focus of all these analyses (see Thompson, 1991). In fact, because all parametric methods (e.g., analysis of variance, regression, multivariate analysis of variance) are special cases of canonical correlation analysis, and one kind of EFA is always implicitly invoked in canonical analysis (Thompson, 1984, pp. 11–13), it can be shown that an EFA is an implicit part of all parametric statistical tests. And as Gorsuch (2003) recently pointed out, "no factor analysis is ever completely exploratory" (p. 143).

ORGANIZATION OF THE BOOK

The book is organized to cover EFA first, followed by CFA. The concepts underlying EFA generally have counterparts in CFA. In fact, a major focus of the book is communicating the linkages between EFA and CFA. In general, if a statistical issue is important in EFA, the same issue is important in CFA. For example, just as EFA pattern and structure coefficients are both important in interpretation when factors are correlated, the same sets of coefficients are important when factors are correlated in CFA.

Chapter 2 covers some foundational statistical topics, such as *score world* versus *area world* statistics, and basic matrix algebra concepts. Chapters 3 through 9 elaborate EFA decisions and analytic choices. Finally, chapters 10 through 12 cover CFA methods.

Each of the chapters concludes with a section called "Major Concepts." These sections are not intended to summarize all the key points of the chapters. Instead, these sections emphasize major concerns that are insufficiently recognized in contemporary practice, or that otherwise might get lost in the details of chapter coverage.

2

FOUNDATIONAL CONCEPTS

A simple heuristic example of exploratory factor analysis (EFA) may make the discussion of some foundational concepts more concrete. Along the way a few elementary ideas involving a type of mathematics called *matrix algebra* will be introduced. It is sometimes jokingly said that matrix algebra invariably induces psychosis. It is true that matrix algebra in some respects is very different from the simple algebra taught in high school. And knowing the mechanics of matrix algebra thankfully is *not* necessary for understanding factor analysis. But we do need to know that some basic *concepts* of matrix algebra (such as when matrices can be multiplied and how division is performed with matrices) are important, even though the mechanics of the process are unimportant.

HEURISTIC EXAMPLE

Table 2.1 presents hypothetical data involving seven students' ratings of me regarding how strongly they agree (9 = *most agree*) versus disagree (1 = *most disagree*) that I am Handsome, Beautiful, Ugly, Brilliant, Smart, and Dumb. Because these variables are directly measured, without applying any weights (e.g., regression β weights) to the scores, these variables are called synonymously *measured* or *observed* variables.

As we shall soon see, scores that are obtained by applying weights to the observed scores are synonymously called *composite, latent,* or *synthetic* variables. All analyses within the general linear model (GLM) yield scores

TABLE 2.1
Heuristic EFA Data

Student/ Statistic	Measured variable					
	Handsome	Beautiful	Ugly	Brilliant	Smart	Dumb
Barbara	6	5	4	8	6	2
Deborah	8	7	2	7	5	3
Jan	9	8	1	9	7	1
Kelly	5	4	5	9	7	1
Murray	4	3	6	9	7	1
Susan	7	6	3	7	5	3
Wendy	3	2	7	7	5	3
Mean	6.00	5.00	4.00	8.00	6.00	2.00
SD	2.16	2.16	2.16	1.00	1.00	1.00

for every participant on at least one composite variable, and often on several composite variables.

When we conduct a factor analysis, we are exploring the relationships among measured variables and trying to determine whether these relationships can be summarized in a smaller number of latent constructs. Several different statistics can be used to summarize the relationships among the variables (e.g., Pearson correlation coefficients, Spearman's rho coefficients). Here we use Pearson product–moment correlation coefficients. Table 2.2 presents the matrix of Pearson correlation coefficients for all the pairwise combinations of these six variables.

The Table 2.1 data set involves relatively few participants ($n = 7$) and relatively few variables ($v = 6$). These features make the data set easy to understand. However, the data and the results are unrealistic. Real EFA problems involve more participants and more variables. Factor analysis is usually less relevant when we have so few variables, because we gain little by summarizing relationships among so few measured variables in a smaller set of latent variables.

TABLE 2.2
Bivariate Correlation Matrix

Variable	Variable					
	Handsome	Beautiful	Ugly	Brilliant	Smart	Dumb
Handsome	1.0	1.0	−1.0	.0	.0	.0
Beautiful	1.0	1.0	−1.0	.0	.0	.0
Ugly	−1.0	−1.0	1.0	.0	.0	.0
Brilliant	.0	.0	.0	1.0	1.0	−1.0
Smart	.0	.0	.0	1.0	1.0	−1.0
Dumb	.0	.0	.0	−1.0	−1.0	1.0

Furthermore, we would rarely confront a correlation matrix, such as that in Table 2.2, in which the correlation coefficients are only −1.0, +1.0, or 0.0. Nor would we tend to see relationships for which the latent constructs are so obvious from a simple examination of the correlation matrix even without performing any statistical analyses.

An examination of the Table 2.2 correlations suggests that two factors (i.e., latent variables) underlie the covariations in the Table 2.1 data. Three patterns are obvious for the heuristic example.

First, one factor involves perceptions of my physical attractiveness. Three variables have r^2 values that are 1.0 (or 100%): Handsome, Beautiful, and Ugly, although scores on Ugly are scaled in the opposite direction from scores on Handsome and Beautiful. Second, another factor involves perceptions of my intellectual prowess. Three variables have r^2 values that are 1.0 (or 100%): Brilliant, Smart, and Dumb, although scores on Dumb are scaled in the opposite direction from scores on Brilliant and Smart. Third, the fact that all pairwise combinations of r^2 values are 0.0 (or 0%) for combinations involving one measured variable from one set (e.g., Handsome) and one measured variable from the other set (e.g., Brilliant) suggests that the two latent variables also are uncorrelated with each other.

TWO WORLDS OF STATISTICS

All statistics can be conceptualized as existing in or describing dynamics of one of two "worlds": the *score* world or the *area* world. All statistics in the score world are in an unsquared metric, as are the scores themselves. For example, if we have data on how many dollars our classmates have on their persons, each measured variable score is in the metric, dollars. Statistics in the score world have implicit superscripts of 1. Example statistics in this world are the mean, the median, the standard deviation (SD), and the correlation coefficient (r).

If the mean is 7, the mean is 7 *dollars*. If the SD is 3, the SD is 3 *dollars*. The weights (e.g., regression β [beta] weights, regression b weights) applied to measured variables to obtain scores on composite variables are also in the score world (or they couldn't be applied to the measured variables!).

The statistics in the area world are in a squared metric. For example, if the scores are measured in dollars, and the variance is 9, the variance (SD^2) is 9 *squared dollars* (not 9 dollars). Many of the statistics in this world have explicit superscripts of 2 (e.g., r^2, R^2) or names that communicate that these statistics are squared values (e.g., the sum of squares, the mean square). Other statistics in this world do not have either superscripts of 2 or names clearly indicating that they are in the score world (e.g., the variance, the covariance).

TABLE 2.3
Unsquared and Squared Factor Weights

Measured variable	Factor		Squared values	
	I	II	I	II
Handsome	1.0	.0	100%	0%
Beautiful	1.0	.0	100%	0%
Ugly	−1.0	.0	100%	0%
Brilliant	.0	1.0	0%	100%
Smart	.0	1.0	0%	100%
Dumb	.0	−1.0	0%	100%

Any statistics we compute as ratios of one area world statistic to another area world statistic are also themselves in the area world. For example, both R^2 and the analysis of variance (ANOVA) effect size η^2 (eta-squared) are computed as:

$$\frac{\text{sum of squares}_{\text{EXPLAINED}}}{\text{sum of squares}_{\text{TOTAL}}}.$$

Reliability coefficients (e.g., Cronbach's α) are also in the area world, because they are computed as ratios of reliable variance or reliable sum of squares divided by total score variance or the total sum of squares. Reliability statistics are in the area world, even though they can be negative (e.g., −.3), and indeed even though they can be less than −1.0 (e.g., −7.0; Thompson, 2003).

Table 2.3 presents the unsquared factor weights (analogous to regression β weights) for the Table 2.2 correlation matrix (and the Table 2.1 data used to create the correlation matrix). Table 2.3 also presents the squared values associated with this analysis. To help readers keep straight which statistics are in the area versus the score world, in many cases I present these statistics as percentages (e.g., $r^2 = 1.0 = 100\%$).

The Table 2.3 results from the factor analysis of the Table 2.2 correlation matrix confirm our expectations. The Table 2.3 matrix (and the Table 2.1 data and the Table 2.2 correlation matrix) suggests that two uncorrelated factors (i.e., my perceived physical attractiveness and my perceived intellectual abilities) underlie the six measured variables.

SOME BASIC MATRIX ALGEBRA CONCEPTS

As noted previously, it is not important to know how to do matrix algebra (which is fortunate, because as noted previously this knowledge can cause severe psychological distress). Readers who desire an understanding of the mechanics of matrix algebra can consult any number of books (e.g., Cooley & Lohnes, 1971, pp. 15–20; Tatsuoka, 1971, chap. 2). But it is important to know some matrix algebra concepts, so that one can have some understanding of what is being done in certain analyses!

Different books use different notations to indicate that a matrix is being discussed. In this book, capital letters in bold are used to represent a matrix (e.g., **X, R**). There are six matrix algebra concepts that must be mastered. These are discussed below.

Rank

In this book we primarily consider matrices that are two-dimensional (i.e., number of rows > 1, and number of columns > 1). Matrix *rank* characterizes how many rows and columns a particular matrix has. When we specify rank, we always report the number of rows first, and the number of columns second, using subscripts. For example, the measured variable scores reported in Table 2.1 could be represented as: $\mathbf{X}_{7 \times 6}$. And the bivariate correlation matrix presented in Table 2.2 is $\mathbf{R}_{6 \times 6}$.

Transpose

For some purposes we can reorganize the data in a matrix such that the columns in the original matrix become the rows in a second matrix, and the rows in the original matrix become the columns in the second matrix. Such a new matrix is called a *transpose*. We designate a transpose by putting a single quotation or tick mark on the matrix. For example, if we have a matrix, $\mathbf{Y}_{3 \times 2}$, the transpose of this matrix is $\mathbf{Y}_{2 \times 3}'$.

Consider the following matrix, $\mathbf{Y}_{3 \times 2}$:

  1 4
  2 5
  3 6.

The transpose of this matrix, $\mathbf{Y}_{2 \times 3}'$, is:

  1 2 3
  4 5 6.

Conformability

In regular (nonmatrix) algebra, any two numbers can be multiplied times each other (e.g., $4 \times 9 = 36$; $2 \times 8 = 16$). In matrix algebra, however, matrices can be multiplied times each other only if a certain condition is met: The number of columns in the left of two matrices being multiplied must equal the number of rows in the right matrix being multiplied. Matrices meeting this condition (and eligible for multiplication) are said to be *conformable*. For example, the Table 2.1 matrix, $X_{7 \times 6}$, can be postmultiplied by the Table 2.2 correlation matrix, $R_{6 \times 6}$, because 6 (columns in X) equals 6 (the number of rows in R).

When we multiply matrices, the rank of the result or product matrices will equal the number of rows in the leftmost matrix, and the number of columns in the product matrix will equal the number of columns in the rightmost matrix. For example, the matrices $Z_{7 \times 6}$ and $W_{6 \times 2}$ are conformable. The product of these two matrices would have a rank of 7 by 2: $Z_{7 \times 6}$ $W_{6 \times 2} = O_{7 \times 2}$.

Symmetric Matrix

Some "square" matrices (number of rows equal number of columns) are *symmetric*, meaning that every *i,j* element of the matrix equals the corresponding *j,i* element of the same matrix. For example, in the Table 2.2 correlation matrix, the *3,1* element in R equals −1.0 equals the *1,3* element in R. Correlation matrices are always symmetric. A related property of symmetric matrices, which are always conformable to themselves, is that a symmetric matrix always equals its own transpose, by definition (i.e., $R_{6 \times 6} = R_{6 \times 6}'$).

Identity Matrix

In regular algebra, any number times 1 equals the original number that was multiplied times 1 (e.g., $9 \times 1 = 9$; $15 \times 1 = 15$). In matrix algebra there is a matrix, called the *identity matrix* (I), which when multiplied times any matrix, equals the initial matrix. For example, let's say we have $Y_{3 \times 2}$:

1 4
2 5
3 6.

For this example, to be conformable the identity matrix must have two rows. The identity matrix is always symmetric. The identity matrix always has ones on the "diagonal" running from the upper left corner down to the lower right corner, and zeroes in all off-diagonal locations. So for this example $I_{2 \times 2}$ would be:

1 0
0 1.

For this example $Y_{3 \times 2} I_{2 \times 2}$ (i.e., Y postmultiplied by I) yields a product matrix that contains exactly the same numbers that are in $Y_{3 \times 2}$ itself. The identity matrix is a critical matrix algebra concept, because as we see next **I** is invoked in order to be able to do matrix division.

Matrix Inverse

For symmetric matrices, one could state a matrix equation such as $\mathbf{R}_{6 \times 6} \mathbf{R}_{6 \times 6}^{-1} = \mathbf{I}_{6 \times 6}$. The matrix, $\mathbf{R}_{6 \times 6}^{-1}$ (i.e., a matrix with a -1 superscript), is called an *inverse*. Not every symmetric matrix has an inverse. If a matrix eligible for inversion (e.g., a symmetric matrix) does not have an inverse for a given data set, the matrix is said to be "ill conditioned."

Some symmetric matrices, such as some correlation matrices, do not have an inverse, because the numbers in the matrix cannot yield an $\mathbf{R}^{-1}$ such that $\mathbf{R} \mathbf{R}^{-1} = \mathbf{I}$. If this happens in factor analysis, this may be due to sample size being too small.

Some factor analyses *require* that an inverse be computed in order to compute the factor solution. When using factor analyses requiring matrix inversion, if a matrix is ill conditioned, the solution simply cannot be computed using the particular technique requiring the inversion. Another technique must be selected, or sample size must be increased.

In regular algebra, division is straightforward. For example, to compute the $F_{CALCULATED}$ in ANOVA or in multiple regression, we compute $F = $ [Sum of Squares$_{EXPLAINED}$ / $df_{EXPLAINED}$] / [Sum of Squares$_{UNEXPLAINED}$ / $df_{UNEXPLAINED}$].

But in matrix algebra the way that *we do division is by postmultiplying the numerator matrix times the inverse of the matrix with which we wish to divide*. For example, the division of $\mathbf{B}_{2 \times 2}$ by $\mathbf{W}_{2 \times 2}$ is accomplished using the formula: $\mathbf{B}_{2 \times 2} \mathbf{W}_{2 \times 2}^{-1} = \mathbf{F}_{2 \times 2}$. (This is actually one basis for obtaining $F_{CALCULATED}$ for a multivariate analysis of variance [MANOVA] and is directly analogous to the univariate ANOVA computation, except for the use of matrix algebra.)

Division is used a lot in univariate statistics. Division is also used a lot in multivariate statistics, including factor analysis. Clearly, access to an inverse is usually critical for most factor analytic applications.

MORE HEURISTIC RESULTS

Throughout the GLM, weights are applied to the scores on the measured variables to obtain scores on composite variables. In the GLM, weights are invoked (a) to compute scores on the latent variables or (b) to interpret what the composite variables represent. Logically, because the weights are used to obtain the scores on the composite variable, the weights must be of some assistance in understanding the nature of the latent variables (i.e., help in interpreting what latent variables represent).

For example, in multiple regression these weights are often called β (*beta*) *weights*. A set of weights for the predictors in regression is called the *regression equation*. These β weights in regression are also sometimes called *standardized regression coefficients* because they are applied to the measured predictor variables in their z-score form (i.e., the measured variables have been transformed so that their means equal zero and their standard deviations and variances equal one). It is actually a misnomer to call the weights "standardized," because the weights are a constant for a given data set. Thus, the weight for the first predictor itself cannot be standardized.

If $\beta_1 = .5$, this weight is applied to the first predictor z scores of everybody in the sample. We cannot apply the formula $z = (\underline{X}_i - \overline{X}) / SD$ to a given *weight*. If everybody is assigned a weight of $\beta_1 = .5$, $SD_{\beta 1} = 0$, then the division is mathematically impermissible. To be technically correct, we should call such weights "weights applied to the standardized measured variables" rather than "standardized weights."

Perhaps we are performing a regression using two predictor variables. Steve has $z_1 = .7$ and $z_2 = .5$. The form of the regression equation is:

$$z_Y \leftarrow \underline{\hat{Y}} = \beta_1(z_1) + \beta_2(z_2).$$

For a given data set, β_1 might equal 2.0, and β_2 might equal -1.2. Note that, as is the case throughout the GLM, weights are *not* necessarily correlation coefficients (see Courville & Thompson, 2001). Thus, as in this example, it is possible that *none* of the weights will lie in the range between -1 and $+1$. For Steve, $\underline{\hat{Y}}_{STEVE} = 2.0(0.7) + -1.2(0.5) = 1.4 + -0.6 = 0.8$. In regression, these $\underline{\hat{Y}}$ scores are the composite, latent, or synthetic variables.

Pattern Coefficients

In factor analysis, *pattern coefficients* are the weights applied to the measured variables to obtain scores on the factor analysis latent variables (called *factor scores*). As noted, these weights are analogous to the β weights in multiple regression. These weights are also analogous to the standardized discriminant function coefficients in descriptive discriminant analysis (Huberty, 1994) and the standardized canonical function coefficients in canonical correlation analysis (Thompson, 1984, 2000a).

The factor pattern coefficients ($\mathbf{P}_{V \times F}$) are computed partly to reexpress the variance represented in the correlation matrix that is being analyzed, and from which the factors (i.e., the sets of pattern coefficients) are extracted. The factors are extracted so that the first factor can reproduce the most variance in the matrix being analyzed, the second factor reproduces the second most variance, and so forth. The ability of one or more factors to reproduce the matrix being analyzed is quantified by the reproduced matrix of associations, such as the reproduced intervariable correlation matrix $(\mathbf{R}_{V \times V}{}^{+})$. This is the "how full is the glass" perspective.

The ability of the factors to reproduce the matrix being analyzed can also be quantified by computing the matrix that is left after a given number of factors have been extracted. This is called the residual matrix of associations, such as the residual intervariable correlation matrix $(\mathbf{R}_{V \times V}{}^{-})$. This is the converse, "how empty is the glass" perspective.

If the factor pattern coefficients perfectly reproduced the matrix of associations (here an intervariable Pearson correlation coefficient matrix), then $\mathbf{R}_{V \times V}{}^{-}$ would consist entirely of zeroes, indicating that there is no information or variance left in this matrix. And if the factor pattern coefficients perfectly reproduced the matrix of associations, then the entries in $\mathbf{R}_{V \times V}{}^{+}$ would exactly match the entries in $\mathbf{R}_{V \times V}$.

The reproduced intervariable correlation matrix $(\mathbf{R}_{V \times V}{}^{+})$ is computed as:

$$\mathbf{P}_{V \times F} \, \mathbf{P}_{F \times V}{}' = \mathbf{R}_{V \times V}{}^{+}.$$

Note that the pattern coefficient matrix and its transpose are conformable, and the product matrix has the rank v by v (the same rank as the original intervariable correlation matrix).

Possible Numbers of Factors

In regression there is only a single "equation" (set of weights, β weights) in a given analysis. In factor analysis the sets of weights (i.e., pattern coefficients) are called *factors* rather than *equations*, mainly just to be inconsistent and to confuse the graduate students. The number of factors worth keeping can range between one and the number of variables.

Let's go back to our example of ratings by seven participants of me as regards six measured variables. If every entry in the intervariable correlation matrix was either a +1 or a −1, the r^2 between every pair of measured variables would be 100%. This would imply that a single factor underlay the ratings. We would extract one factor $(\mathbf{P}_{6 \times 1})$ consisting exclusively of negative or plus ones. And this single factor would perfectly reproduce the original $\mathbf{R}_{6 \times 6}$ matrix.

Technically, there would be five additional factors, each consisting exclusively of zeroes, meaning that they each contain and reproduce no information. But we would never bother with such factors. Indeed, we usually do not bother even with factors that reproduce some, but only a little information, if we deem the amount of information or variability reproduced to not be noteworthy.

At the other extreme, if somehow all the ratings were perfectly uncorrelated, then every off-diagonal element of $\mathbf{R}_{6 \times 6}$ would be a zero (and $\mathbf{R}_{6 \times 6} = \mathbf{I}_{6 \times 6}$). Now, no two variables could be combined to create a factor (i.e., each measured variable would define its own factor). There would be six factors. Each factor would have one +1 value, and the remaining five entries would be zeroes.

At this second extreme (perfectly uncorrelated measured variables), $\mathbf{R}_{V \times V}^{+}$ would exactly match the entries in $\mathbf{R}_{V \times V}$. In fact, whenever we extract all possible factors (i.e., the number of factors equals the number of measured variables), the pattern coefficients will perfectly reproduce the original matrix of associations being analyzed.

Factor Structure Coefficients

All GLM analyses are correlational and yield weights (e.g., factor pattern coefficients) that are applied to measured variables to obtain scores on composite variables. These weights are sometimes correlation coefficients, and sometimes they are not (cf. Courville & Thompson, 2001; Thompson, 1984, pp. 22–23).

Researchers can also compute bivariate correlation coefficients between measured variables and their composite variables (e.g., regression $\hat{Y}$ scores, factor scores, discriminant function scores). These correlations across various GLM analyses are always called *structure coefficients*.

Throughout the GLM consulting the structure coefficients is almost always essential to correctly interpreting results. For example, in regression Courville and Thompson (2001) documented the misinterpretations that can occur when only β weights are consulted in interpretation (also see Cooley & Lohnes, 1971, p. 55; Dunlap & Landis, 1998; Thompson & Borrello, 1985).

For descriptive discriminant analysis, Huberty (1994) noted that "construct definition and structure dimension [and not hit rates] constitute the *focus* [italics added] of a descriptive discriminant analysis" (p. 206). The importance of structure coefficients in canonical correlation analysis is widely recognized (cf. Cohen & Cohen, 1983, p. 456; Levine, 1977, p. 20). For example, Meredith (1964) noted, "If the variables within each set are moderately intercorrelated the possibility of interpreting the canonical variates by inspection of the appropriate regression weights [function coefficients] is practically nil" (p. 55).

Structure coefficients are equally important in factor analysis (cf. Graham, Guthrie, & Thompson, 2003; Thompson, 1997). Thus, Gorsuch (1983) emphasized that a *"basic* [italics added] matrix for interpreting the factors is the factor structure" (p. 207).

In the GLM, *sometimes* the weights are correlation coefficients. For example, in regression, if the predictor variables are perfectly uncorrelated, the β weight for a given predictor equals that variable's correlation with the outcome variable (Y_i).

In factor analysis, the structure coefficients ($\mathbf{S}_{V \times F}$) can be computed using the equation

$$\mathbf{P}_{V \times F} \, \mathbf{R}_{F \times F} = \mathbf{S}_{V \times F},$$

where $\mathbf{R}_{F \times F}$ is the interfactor correlation matrix (not the v by v intervariable correlation matrix). When the factors are perfectly uncorrelated, by definition $\mathbf{R}_{F \times F} = \mathbf{I}_{F \times F}$, and therefore $\mathbf{P}_{V \times F} = \mathbf{S}_{V \times F}$.

In other words, whenever the factors are perfectly uncorrelated the pattern coefficients are also structure coefficients (and therefore must fall within the range −1 to +1), and in this case are more appropriately termed *pattern/structure coefficients*. And when factors are first extracted, the factors are *always* perfectly uncorrelated.

"Loadings"

It is not uncommon in factor analytic reports to see authors referring to coefficients as "loadings." The problem is that this term is inherently ambiguous. In some reports authors refer to pattern coefficients as "loadings." In other reports authors refer to structure coefficients as "loadings." In some reports it is impossible to discern what coefficients the authors are describing.

And in some reports the pattern and the structure coefficients are not equal but authors refer to both sets of coefficients as "loadings." These ambiguities cause confusion that impedes clear scientific dialogue. Therefore, some journals (see Thompson & Daniel, 1996) formally discourage the use of this vague colloquialism. For the same reasons, the term *loading* is not used here.

Communality Coefficients

For the Table 2.3 factors, because the factors are perfectly uncorrelated, the unsquared coefficients reproduced in Table 2.4 are pattern/structure coefficients. Recall that Pearson correlation coefficients (a) characterize the linear relationship between two intervally scaled variables but (b) the coefficients themselves are *not* intervally scaled. That is, an r of 1.0 is *not* twice as large as an r of 0.5.

To make the correlations intervally scaled, so that we can say how much larger one r is versus another r, we must first square the rs before we compare them. Thus, the r of 1.0 is four times larger than the r of 0.5,

TABLE 2.4
Pattern/Structure Communality Coefficients and Eigenvalues

Measured variable	Factor I	Factor II	h^2
Handsome	1.0	.0	100%
Beautiful	1.0	.0	100%
Ugly	−1.0	.0	100%
Brilliant	.0	1.0	100%
Smart	.0	1.0	100%
Dumb	.0	−1.0	100%
Sum of squared column values	3.0	3.0	

Note. The eigenvalues (3.0 and 3.0) are in a squared, area-world metric, even though they are not presented as percentages.

because $r^2 = 1.0^2 = 100\%$ is four times larger than $r^2 = 0.5^2 = 25\%$. Because the Table 2.4 structure coefficients are correlation coefficients, they too must be squared before they are compared.

One set of manipulations involves taking the squared structure coefficients for the uncorrelated factors (the formula is different for correlated factors) and adding the squared values along the rows across the columns (e.g., 100% + 0% = 100%). The resulting coefficients, called *communality coefficients* (h^2), are area-world statistics. Because these factors are uncorrelated, the variance each measured variable shares with a factor is unique or nonoverlapping, and thus the h^2 value for a measured variable indicates *how much of the variance in a measured variable the factors as a set can reproduce*.

If a measured variable had a communality coefficient close to 0%, this would mean that this variable is not being represented within the factors. If the researcher desires the variable to be represented in the factors, it may be necessary to extract additional factors.

An equally valuable perspective takes a converse view of the interpretation of the communality coefficients. The h^2 value for a measured variable reflects *how much of the variance of a given measured variable was useful in delineating the factors as a set*.

There is yet another perspective from which communality coefficients can be viewed. The communality coefficient for a variable is a "lower bound" estimate of the reliability of the scores on the variable. This means that if a variable has an h^2 of 50%, we believe that the reliability of the scores on the variable is no lower than .5 (and, of course, may be higher).

In some unusual cases communality coefficients greater than 100% occur. These are called Heywood cases (Heywood, 1931). Such solutions are statistically "inadmissible," and they indicate fundamental problems similar to estimating variances to be negative.

Table 2.4 presents the communality coefficients for each of the six measured variables in the heuristic example. Each of these is 100%. This reflects the fact that for the present example, if we reexpress the data presented in Table 2.1 in the form of scores on the two factors presented in Table 2.3, we lose no information whatsoever!

Eigenvalues

For the Table 2.3 factors, if we add the squared structure coefficients down the columns across the rows, we get a very important set of squared, area-world statistics called *eigenvalues* (λ). As is the case elsewhere in statistics, important concepts are given multiple names, to confuse everybody; naturally it is the best use of effort to be confusing to focus such attention primarily on the most important concepts, by giving them the most synonymous names. Another synonymous term for eigenvalues is *characteristic roots.*

The eigenvalues for this simple heuristic example are 3.0 and 3.0, as reported in Table 2.4. (Normally eigenvalues descend in value, but for this heuristic example the first two eigenvalues were exactly equal.) Eigenvalues are always indices of the amount of information represented in some multivariate result. In some multivariate analyses eigenvalues are multivariate squared correlation coefficients. Clearly, given that the first two eigenvalues associated with the factor analysis were both greater than 1.0, in EFA eigenvalues are *not* squared correlation coefficients.

The following four statements apply to eigenvalues in an EFA:

1. The number of eigenvalues equals the number of measured variables being analyzed.
2. The sum of the eigenvalues equals the number of measured variables.
3. An eigenvalue divided by the number of measured variables indicates the proportion of information in the matrix of associations being analyzed that a given factor reproduces.
4. The sum of the eigenvalues for the extracted factors divided by the number of measured variables indicates the proportion of the information in the matrix being analyzed that the factors as a set reproduce.

For the current example the number of measured variables is six. Therefore, there are six eigenvalues associated with the Table 2.2 correlation matrix. Because the sum of the eigenvalues for the current example is 6.0, given that the first two eigenvalues are 3.0 and 3.0, the remaining eigenvalues must be 0.0, 0.0, 0.0, and 0.0.

Because 3.0 / 6 equals 0.5, the first eigenvalue indicates that Factor I reproduces 0.5 (or 50%) of the information contained in the Table 2.2

correlation matrix. The second eigenvalue indicates that Factor II also reproduces 0.5 (or 50%) of the information contained in the Table 2.2 correlation matrix.

The sum of the first two eigenvalues (i.e., 6.0) divided by the number of measured variables (6) indicates that 1.0 (or 100%) of the information contained in the Table 2.2 correlation matrix is reproduced by the factors as a set. This result is consistent with the finding that all the communality coefficients were 100%. These various calculations are routinely provided by statistical packages such as SPSS.

Sequential Factor Extraction

In reality all possible factors, equal to the number of variables, are computed at one time. However, researchers rarely, if ever, retain all possible factors, because many potential factors will be deemed trivial. So it is useful to think of the factors as if they were extracted from the matrix of associations (e.g., the bivariate correlation matrix) one at a time, in sequence. And after each extraction a judgment can be made as to whether to extract additional factors.

For the Table 2.1 data, the first factor (i.e., $\mathbf{P}_{6 \times 1}$) that would be extracted is Factor I, presented in Table 2.4. The variance/covariance that can be reproduced ($R_{6 \times 6}^{+}$) is computed for uncorrelated factors by multiplying the pattern matrix ($\mathbf{P}_{6 \times 1}$) times its transpose ($\mathbf{P}_{6 \times 1}'$):

$$
\begin{bmatrix}
 & \text{I} \\
\text{HANDSOME} & 1.0 \\
\text{BEAUTIFUL} & 1.0 \\
\text{UGLY} & -1.0 \\
\text{BRILLIANT} & .0 \\
\text{SMART} & .0 \\
\text{DUMB} & .0
\end{bmatrix}_{6 \times 1}
\begin{bmatrix}
 & \text{HAND} & \text{BEAU} & \text{UGLY} & \text{BRIL} & \text{SMAR} & \text{DUMB} \\
\text{I} & 1.0 & 1.0 & -1.0 & .0 & .0 & .0
\end{bmatrix}_{1 \times 6}
$$

For these data, the bottom triangle of the symmetric *reproduced* correlation ($R_{6 \times 6}^{+}$) is:

$$
\begin{bmatrix}
 & \text{HANDSOME} & \text{BEAUTIFUL} & \text{UGLY} & \text{BRILLIANT} & \text{SMART} & \text{DUMB} \\
\text{HANDSOME} & 1.0 \\
\text{BEAUTIFUL} & 1.0 & 1.0 \\
\text{UGLY} & -1.0 & -1.0 & 1.0 \\
\text{BRILLIANT} & .0 & .0 & .0 & .0 \\
\text{SMART} & .0 & .0 & .0 & .0 & .0 \\
\text{DUMB} & .0 & .0 & .0 & .0 & .0 & .0
\end{bmatrix}_{6 \times 6}
$$

The *residual* variance/covariance ($R_{6\times6}^-$) is computed by subtracting $R_{6\times6}^+$ from $R_{6\times6}$. For these data the result ($R_{6\times6}^-$) is:

$$
\begin{bmatrix}
 & \text{HANDSOME} & \text{BEAUTIFUL} & \text{UGLY} & \text{BRILLIANT} & \text{SMART} & \text{DUMB} \\
\text{HANDSOME} & .0 \\
\text{BEAUTIFUL} & .0 & .0 \\
\text{UGLY} & .0 & .0 & .0 \\
\text{BRILLIANT} & .0 & .0 & .0 & 1.0 \\
\text{SMART} & .0 & .0 & .0 & 1.0 & 1.0 \\
\text{DUMB} & .0 & .0 & .0 & -1.0 & -1.0 & 1.0
\end{bmatrix}_{6\times6}
$$

The next factor can then be extracted from the residual matrix of association. Because all variance reproducible by the first factor has been removed, the first and the second factors will *always* be "orthogonal," or uncorrelated. The result would be the two factors presented in Table 2.4.

The variance/covariance that can be *reproduced* ($R_{6\times6}^+$) is again computed by multiplying the pattern matrix times its transpose:

$$
\begin{bmatrix}
 & \text{I} & \\
\text{HANDSOME} & 1.0 & .0 \\
\text{BEAUTIFUL} & 1.0 & .0 \\
\text{UGLY} & -1.0 & .0 \\
\text{BRILLIANT} & .0 & 1.0 \\
\text{SMART} & .0 & 1.0 \\
\text{DUMB} & .0 & -1.0
\end{bmatrix}_{6\times2}
\begin{bmatrix}
 & \text{HAND} & \text{BEAU} & \text{UGLY} & \text{BRIL} & \text{SMAR} & \text{DUMB} \\
\text{I} & 1.0 & 1.0 & -1.0 & .0 & .0 & .0 \\
\text{II} & .0 & .0 & .0 & 1.0 & 1.0 & -1.0
\end{bmatrix}_{2\times6}
$$

Now the bottom triangle of the *reproduced* correlation ($R_{6\times6}^+$) is:

$$
\begin{bmatrix}
 & \text{HANDSOME} & \text{BEAUTIFUL} & \text{UGLY} & \text{BRILLIANT} & \text{SMART} & \text{DUMB} \\
\text{HANDSOME} & 1.0 \\
\text{BEAUTIFUL} & 1.0 & 1.0 \\
\text{UGLY} & -1.0 & -1.0 & 1.0 \\
\text{BRILLIANT} & .0 & .0 & .0 & 1.0 \\
\text{SMART} & .0 & .0 & .0 & 1.0 & 1.0 \\
\text{DUMB} & .0 & .0 & .0 & -1.0 & -1.0 & 1.0
\end{bmatrix}_{6\times6}
$$

The *residual* variance/covariance ($R_{6\times6}^-$) is again computed by subtracting $R_{6\times6}^+$ from $R_{6\times6}$. For these data the result ($R_{6\times6}^-$) is:

$$
\begin{bmatrix}
 & \text{HANDSOME} & \text{BEAUTIFUL} & \text{UGLY} & \text{BRILLIANT} & \text{SMART} & \text{DUMB} \\
\text{HANDSOME} & .0 & & & & & \\
\text{BEAUTIFUL} & .0 & .0 & & & & \\
\text{UGLY} & .0 & .0 & .0 & & & \\
\text{BRILLIANT} & .0 & .0 & .0 & .0 & & \\
\text{SMART} & .0 & .0 & .0 & .0 & .0 & \\
\text{DUMB} & .0 & .0 & .0 & .0 & .0 & .0
\end{bmatrix}_{6 \times 6}
$$

This result reflects the fact that for these heuristic data the two factors àlone reproduce all the information contained within the original Table 2.2 correlation matrix.

SAMPLE SIZE CONSIDERATIONS IN EXPLORATORY FACTOR ANALYSIS

Sample size affects the precision of all statistical estimates, including those made in EFA. Various researchers have proposed rules of thumb for sample size minimums that are a function of the ratio of the number of people to the number of measured variables. The recommended ratios often are within the range of 10 to 20 people per measured variable. Gorsuch (1983) suggested that "an absolute minimum ratio is five individuals to every variable, but not less than 100 individuals for any analysis" (p. 332).

However, some Monte Carlo simulation research (Guadagnoli & Velicer, 1988) suggests that the most critical issue is how saturated the factors are by the measured variables. They suggested that replicable factors tend to be estimated if:

1. factors are each defined by four or more measured variables with structure coefficients each greater than |.6|, regardless of sample size; or
2. factors are each defined with 10 or more structure coefficients each around |.4|, if sample size is greater than 150; or
3. sample size is at least 300.

MacCallum, Widaman, Zhang, and Hong (1999) found that sample sizes as low as 60 accurately reproduced population pattern coefficients if communalities were all .60 or greater. If h^2 values were around .50, sample sizes of 100 to 200 were required.

Of course, when it comes to mathematically complex analyses such as EFA, more is always better. This principle is particularly relevant, because one may not know prior to conducting the analysis what the saturation pattern of the factors will be, and so rules revolving around saturation patterns are of limited use at the time when we actually make our sample size decisions.

MAJOR CONCEPTS

Factor analyses typically involve dozens or even hundreds of variables, and so the matrices of association being analyzed (e.g., Pearson product–moment correlation matrices) have huge numbers of coefficients, and the matrix entries (unlike the heuristic example) are not limited to +1's, −1's, and 0's. The patterns underlying such matrices must be explored empirically, and not merely by subjective inspection, because no person is sufficiently bright to apprehend the structure underlying so many different coefficients.

Factor analysis requires the use of matrix algebra, a form of algebra not typically encountered within education prior to graduate school. Matrix algebra has some unique features, such as rules regarding what matrices can be multiplied by each other (i.e., conformability). And as readers learned in high school, in conventional algebra the order of terms being multiplied does not affect the calculated result (e.g., $3 \times 2 = 2 \times 3 = 6$). However, in matrix algebra the order of terms may affect the result. For example, assuming both $\mathbf{X} \mathbf{Y}$ and $\mathbf{Y} \mathbf{X}$ are conformable, the two multiplications of the same two matrices, $\mathbf{X}$ and $\mathbf{Y}$, may yield different products!

Division in matrix algebra is also done differently. In matrix algebra, to divide we must first solve for the inverse of the matrix with which we wish to divide. Then the division is accomplished by postmultiplying the matrix we wish to divide by the inverse of the divisor matrix. For example, we divide the entries in the matrix $\mathbf{Z}$ by $\mathbf{R}$ first finding the inverse of $\mathbf{R}$ (i.e., $\mathbf{R}'$), and then multiplying $\mathbf{Z}$ by $\mathbf{R}'$.

It was also noted that all analyses within the GLM (e.g., t tests, ANOVA, r, regression, MANOVA, descriptive discriminant analysis, canonical correlation analysis) apply weights to the measured variables to estimate scores on latent or composite variables (e.g., factor scores in factor analysis). Both the weights and the structure coefficients (i.e., correlations of measured variables with composite variables) are usually important in result interpretation (see Courville & Thompson, 2001).

3

EXPLORATORY FACTOR ANALYSIS
DECISION SEQUENCE

Exploratory factor analysis (EFA) in actuality involves a linear sequence of decisions each involving a menu of several available choices. Given that there are five major decisions, with many choices at each decision point, the number of different analysis combinations that are available to the researcher is quite large. These five decisions, discussed in more detail in the remainder of the present chapter, address the following questions:

1. Which matrix of association coefficients should be analyzed?
2. How many factors should be extracted?
3. Which method should be used to extract the factors?
4. How should the factors be rotated?
5. How should factor scores be computed if factor scores are of interest?

In EFA, using more than one set of analytic choices is usually best practice. As Gorsuch (1983) wisely suggested, "Factor the data by several different analytic procedures and hold sacred only those factors that appear across all the procedures used" (p. 330).

This chapter presents the EFA decision sequence. Not every choice at each decision point is described. But the most common selections are presented. And in addition to covering the most commonly encountered choices, the presentation is sufficiently broad to give the reader a flavor of

TABLE 3.1
SPSS Variable Names and Variable Labels for the Appendix A Data

Variable	SPSS variable label
PER1	'1 Willingness to help users'
PER2	'1 Giving users individual attention'
PER3	'1 Employees who deal with users in a caring fashion'
PER4	'1 Employees who are consistently courteous'
PER5	'2 A haven for quiet and solitude'
PER6	'2 A meditative place'
PER7	'2 A contemplative environment'
PER8	'2 Space that facilitates quiet study'
PER9	'3 Comprehensive print collections'
PER10	'3 Complete runs of journal titles'
PER11	'3 Interdisciplinary library needs being addressed'
PER12	'3 Timely document delivery/interlibrary loan'

the complexity of the choices. Exploratory factor analysis, to put it mildly, is not any one single analysis—EFA encompasses a broad family of choices.

The small, but real, data set presented in Appendix A will be used to make this discussion concrete. The data were randomly sampled from the LibQUAL+^{TM} study of user perceptions of service quality at academic libraries in the United States and Canada (cf. Cook & Thompson, 2001; Thompson, Cook, & Heath, 2001; Thompson, Cook, & Thompson, 2002). Appendix A presents the ratings provided by 100 graduate students and 100 faculty from one LibQUAL+^{TM} study. Table 3.1 presents the 12 measured variables sampled from the data set.

MATRIX OF ASSOCIATION COEFFICIENTS

The scores on measured variables are not the actual basis for EFA. Instead, these data are used to compute some matrix of bivariate associations among the measured variables. It is the matrix of associations that is analyzed within EFA. Indeed, given the matrix of associations (e.g., $\mathbf{R}_{11 \times 11}$) for a data set, all the steps of the factor analysis (except for the computation of factor scores) can be accomplished even without access to original data (e.g., $\mathbf{X}_{100 \times 11}$).

Because the factors are extracted from some matrix of associations, the factors themselves are sensitive only to those dynamics captured in a given statistic measuring bivariate relationships. If the chosen statistic only evaluates the orderings created by the measured variables, ignoring distances, then the factors will also be exclusively order-driven. If the statistic chosen to characterize association is influenced by both relationships and score variabilities, then the factors in turn will also be a function of all these dynamics. And for the same data set, different factors may emerge depending on the selection of the matrix of associations to be analyzed.

The Pearson product–moment bivariate correlation matrix is the matrix of associations most commonly used in EFA. In fact, in most statistical packages the default (used unless the user overrides the default choice) matrix of associations used in EFA is the Pearson correlation matrix.

But there are quite a number of other available choices. Different statistics that characterize relationships are sensitive to different aspects of the data. Different relationship statistics also presume that different levels of scale underlie the data.

For example, the Pearson r requires that data must be intervally scaled. Spearman's rho, on the other hand, presumes only that data are at least ordinally scaled. But if data are intervally scaled, the researcher could use either Pearson or Spearman coefficients as the basis of the analysis. In fact, Spearman's rho is Pearson's r between two variables once interval data have been converted into ranks with no ties.

For example, presume a data set with interval scores of three people on two variables:

Participant	X	Y
Barbara	1	1
Deborah	2	2
Jan	3	96

The Pearson correlation coefficient (r_{XY}) can be computed as:

$$r_{XY} = COV_{XY} / (SD_X * SD_Y),$$

where the covariance of X and Y is computed as:

$$COV_{XY} = (\Sigma \ (X_i - \overline{X})(Y_i - \overline{Y})) \ / \ n - 1,$$

or equivalently (because $x_i = X_i - \overline{X}$ and $y_i = Y_i - \overline{Y}$) as:

$$COV_{XY} = (\Sigma \ xy_i) \ / \ n \ - 1.$$

For these initial data, we obtain:

Participant	X	$\overline{X}$	x	Y	$\overline{Y}$	y	xy
Barbara	1	2.0	−1.0	1	33.0	−32.0	32.0
Deborah	2	2.0	0.0	2	33.0	−31.0	0.0
Jan	3	2.0	1.0	96	33.0	63.0	63.0
Sum	6.00			99.00			95.00
Mean	2.00			33.00			
SD	1.00			54.56.			

So the covariance is 95.00 / 2 = 47.50. And the Pearson r equals:

$$47.50 / (1.00 \times 54.56)$$
$$47.50 / 54.56 = .87.$$

If these intervally scaled data are converted into ranks, the only change is that Jan's score of 96 on Y becomes 3 when the scores are all converted into ranks. Now we can compute the Spearman's rho using the Pearson r formula:

Participant	X	$\overline{X}$	x	Y	$\overline{Y}$	y	xy
Barbara	1	2.0	−1.0	1	2.0	−1.0	1.0
Deborah	2	2.0	0.0	2	2.0	0.0	0.0
Jan	3	2.0	1.0	3	2.0	1.0	1.0
Sum	6.00			6.00			2.00
Mean	2.00			2.00			
SD	1.00			1.00			

The covariance is 2.00 / 2 = 1.00. And the Pearson r equals:

$$1.00 / (1.00 \times 1.00)$$
$$1.00 / 1.00 = 1.00.$$

In effect, whether computed on ordinal or on interval data, Spearman's rho addresses the question, "Do the two variables order the people in exactly the same order?" Pearson's r evaluates this question also, but takes into account as well the distances between the ordered scores. Spearman's rho presumes that no such information is present in data (or ignores this information), which is why rho can be used when both variables are ordinally scaled.

These two matrices of association are two of the choices that researchers can use in extracting factors. But a third choice is the matrix of covariances itself. In fact, the covariance matrix is the most common choice as the matrix of associations investigated in confirmatory factor analysis (CFA).

From the formula $r_{XY} = COV_{XY} / (SD_X * SD_Y)$, we can see that r_{XY} will equal COV_{XY} if either (a) r (or the covariance) is zero, or (b) the two standard deviations are reciprocals of each other (e.g., 1 and 1, .5 and 2). In all other cases, $r_{XY} \neq / COV_{XY}$.

In many contexts the covariance is used primarily as an intermediate calculation in obtaining the correlation coefficient, and not to describe relationship or association. The covariance is used infrequently because, unlike r, the covariance does not have a definitive range of possible values. For example, consider the following combinations of r and SD values:

$$
\begin{array}{ccccccc}
r_{XY} & * & SD_X & * & SD_Y & = & COV_{XY} \\
1.00 & * & 100 & * & 100 & = & 10000 \\
0.50 & * & 100 & * & 200 & = & 10000 \\
0.01 & * & 1000 & * & 1000 & = & 10000.
\end{array}
$$

Clearly the covariance can equal a given value under a wide array of circumstances!

The covariance is jointly influenced by three aspects of the two variables: the correlation between the two variables, the variability of the first variable, and the variability of the second variable. Therefore, when exploratory factors are extracted from a covariance matrix, some factors may be a function of correlations, whereas others may be more a function of score spreadoutness. Sometimes we want our factors to be sensitive to an array of aspects of the scores. But at other times we may prefer all the factors to be sensitive only to a single aspect of our data.

NUMBER OF FACTORS

A critical decision in any EFA is determining how many factors to retain. There are numerous strategies for making this decision. In general, several strategies should be used with the hope that different approaches to making this decision will corroborate each other. Although empirical evidence can inform this judgment, these decisions are in the final analysis matters of exactly that: judgment. Zwick and Velicer (1986) presented what is the most authoritative empirical evaluation of how well these various strategies work.

Statistical Significance Tests

Statistical significance tests due to Bartlett (1950) can be used in either of two ways. First, the correlation matrix can be tested to evaluate whether the matrix is an identity matrix. If this null hypothesis that the correlation matrix is an identity matrix cannot be rejected, then factors cannot sensibly be extracted from the matrix. Second, after each factor is extracted, with factors being extracted successively so as to contain progressively less and less information or variance, the residual correlation matrix can be analyzed to evaluate whether any information remains (i.e., whether the residual matrix is essentially an identity matrix).

The problem with these applications is the same general problem that applies in all statistical significance tests (see Thompson, 1996). Statistical significance is driven in large part by sample size. Because researchers generally use EFA only with reasonably large samples, even trivial correlations or factors will be adjudicated statistically significant. Therefore, this approach is not terribly useful.

This does not mean that a researcher should factor analyze a correlation matrix for which the test of the observed correlation matrix is not statistically significant. But, as a practical matter, given any even only approximately reasonable sample size, the null hypothesis that the correlation matrix is an identity matrix will always be rejected.

Eigenvalue Greater Than 1.0 Rule

In 1954 Guttman reasoned that noteworthy factors should have eigenvalues greater than 1.0. Sometimes this logic is attributed to Kaiser, and called the "K1" rule. This is the default decision-making strategy for determining the number of factors in most statistical packages. Consequently, this rule is commonly encountered in the literature, either explicitly when authors justify their extraction decisions or implicitly when researchers less thoughtfully rely on the default decisions built into common software.

The logic underlying this rule is reasonable. Measured and composite variables are separate classes of variables. Factors, by definition, are latent constructs created as aggregates of measured variables and so should consist of more than a single measured variable. If a factor consisted of a single measured variable, even if that measured variable had a pattern/structure coefficient of 1.0 (or −1.0) and all other variables on that factor had pattern/structure coefficients of .0, the factor would have an eigenvalue of 1.0. Thus, it seems logical that noteworthy factors (i.e., constructs representing aggregates of measured variables) should have eigenvalues greater than 1.0.

However, it is very important for the researcher to exercise some judgment in using this strategy to determine the number of factors to extract or retain. Eigenvalues, like all sample statistics, have some sampling error. Therefore, informed by theory and previous related EFA research, a thoughtful researcher may extract a factor with an eigenvalue of .999 or .950, or conversely not retain a factor with an eigenvalue of 1.005 or 1.100.

Scree Test

Cattell (1966b) proposed a graphical test for determining the number of factors. As noted previously, eigenvalues characterize the amount of information represented within a given factor. Factors have successively smaller eigenvalues.

Cattell based his method on the concept of *scree*. Scree is the rubble of loose rock and boulders not solidly attached to mountains that collects at the feet of the mountains. Cattell thought of solid, big, intact mountains as being analogous to solid, noteworthy factors that researchers should

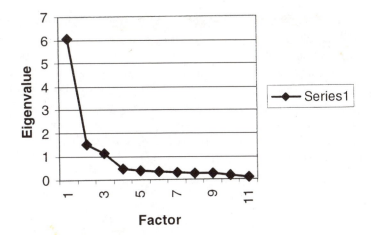

Figure 3.1. Scree plot for first 100 cases and 11 variables presented in Appendix A.

recognize and retain. Trivial factors, however, are analogous to scree, and should be left behind in the extraction process.

A scree plot graphs eigenvalue magnitudes on the vertical access, with eigenvalue numbers constituting the horizontal axis. The eigenvalues are plotted as asterisks within the graph, and successive values are connected by a line. Factor extraction should be stopped at the point where there is an "elbow," or leveling of the plot. This visual approach, not invoking statistical significance, is sometimes called a "pencil test," because a pencil can be laid on the rightmost portion of the relevant graphic to determine where the elbow or flattening seems to occur.

Statistical packages provide the scree plot on request. Of course, because this strategy requires judgment, different researchers given the same plot may disagree regarding what number of factors should be retained.

Figure 3.1 presents the scree plot associated with the analysis of the first 100 cases of data for the first 11 variables listed in Appendix A. These data are perceptions of 100 graduate students regarding library service quality on the first 11 variables listed in Table 3.1. This plot suggests that three factors should be extracted. Empirical variations on this "visual scree" test use regression analyses to evaluate objectively where the elbow in the plot occurs (Nasser, Benson, & Wisenbaker, 2002).

Inspection of the Residual Correlation Matrix

As noted previously, as more factors are extracted the entries in the residual correlation matrix ($\mathbf{R}_{V \times V}$) approach zeroes. If all possible factors are extracted, then the residual matrix will always consist only of zeroes.

Logically, another approach to determining the number of noteworthy factors involves examination of the residual matrix as successive factors are extracted. Computer packages provide the residual matrix on request. And some packages count the number of entries in one triangle of the residual correlation matrix that are greater than |.05| to help inform this evaluation. Note that this criterion has nothing to do with $p < .05$ and that the cutoff value is somewhat arbitrary. Researchers may invoke other criteria, and might elect to use larger cutoffs when sample sizes are small and the standard errors of the correlation coefficients are larger.

Parallel Analysis

Horn (1965) proposed a strategy called *parallel analysis* that appears to be among the best methods for deciding how many factors to extract or retain (Zwick & Velicer, 1986). The strategy involves taking the actual scores on the measured variables and creating a random score matrix of exactly the same rank as the actual data with scores of the same type represented in the data set. There are various approaches to this strategy.

The focus of the method is to create eigenvalues that take into account the sampling error that influences a given set of measured variables. If scores are randomly ordered, the related correlation matrix will approximate an identity matrix, and the eigenvalues of that matrix will fluctuate around 1.0 as a function of sampling error. The range of fluctuations will tend to be narrower as the sample size is larger and the number of measured variables is smaller.

If one presumes that actual data are approximately normally distributed, then the short SPSS syntax presented by Thompson and Daniel (1996) may be used to conduct the analysis. This program creates randomly ordered scores consisting of the same integer values (e.g., 1, 2, 3, 4, or 5) represented in some data sets such as Likert scale data.

Other approaches use regression equations published in prior simulation studies to predict the eigenvalues that would be produced from the correlation matrices for randomly ordered data having a certain rank. Montanelli and Humphreys (1976) produced one such equation that uses as input only the actual sample size and the number of measured variables. Lautenschlager, Lance, and Flaherty (1989) published an alternative formula that uses these two predictors but also uses the ratio of variables-to-participants as another predictor variable.

A more sophisticated approach exactly honors not only the types of possible values contained in the data set but also the exact values (and their descriptive statistics including shape statistics such as skewness) in the real data set. Table 3.2 presents the first 25 scores from Appendix A on the variables PER1, PER2, PER5, and PER6. The data on these variables

TABLE 3.2
Variables Randomly Sorted for Parallel Analysis

ID	Original data				Randomly sorted data			
	PER1	PER2	PER5	PER6	RAN1	RAN2	RAN5	RAN6
1	8	7	3	2	9	5	3	6
2	5	7	4	5	8	9	7	4
3	6	5	5	3	7	5	5	5
4	5	5	4	4	5	7	4	4
5	5	5	5	4	5	5	4	7
6	7	7	7	8	5	5	5	3
7	8	8	6	7	9	9	5	3
8	7	7	5	3	9	7	6	5
9	1	3	1	1	8	9	2	8
10	9	9	5	7	8	3	5	7
11	9	9	7	9	8	8	1	9
12	4	3	6	5	7	3	7	6
13	6	6	7	7	1	7	7	1
14	7	7	2	3	8	9	5	3
15	9	9	7	5	7	9	7	5
16	9	9	5	5	9	7	5	7
17	8	8	7	6	9	7	9	3
18	8	5	4	8	7	7	6	6
19	9	9	6	7	6	6	4	6
20	8	7	5	6	4	8	7	2
21	9	9	9	9	8	8	6	6
22	7	7	7	6	7	7	7	5
23	7	8	5	6	6	9	5	9
24	8	9	5	3	8	8	6	8
25	8	8	6	6	9	9	5	7
Mean	7.08	7.04	5.32	5.40	7.08	7.04	5.32	5.40
SD	1.93	1.84	1.75	2.14	1.93	1.84	1.75	2.14

were independently randomly ordered to create the scores on variables RAN1, RAN2, RAN5, and RAN6.

Note that the means and standard deviations on related actual and randomly ordered data match exactly (e.g., $\overline{X}_{PER1} = \overline{X}_{RAN1} = 7.08$; $SD_{PER1} = SD_{RAN1} = 1.93$). Other descriptive statistics, such as coefficients of skewness and kurtosis, would also match for each actual and randomly ordered variable pair.

However, randomly ordered scores should have bivariate correlations approaching zero with eigenvalues all fluctuating around one. Table 3.3 presents the correlation matrix for the actual data presented in Table 3.2, and correlation coefficients for the same data after independent random ordering of the scores on each variable. Note that the correlation coefficients for the randomly ordered variables are closer to zero than the correlations among the actual scores.

TABLE 3.3
Bivariate Correlations For Actual and Randomly Ordered Data

Variable	PER1/ RAN1	PER2/ RAN2	PER5/ RAN5	PER6/ RAN6
PER1/RAN1		.15	−.13	.41
PER2/RAN2	.87		.01	−.02
PER5/RAN5	.51	.46		−.55
PER6/RAN6	.54	.43	.73	

Note. The coefficients above the diagonal are correlations among the randomly ordered data. The coefficients below the diagonal are correlations among the actual data in its original order, as presented in Table 3.2.

TABLE 3.4
Eigenvalues for Actual and Randomly Ordered Data

Number	Actual data eigenvalue	Random order eigenvalue
1	2.776	1.756
2	0.828	1.082
3	0.276	0.797
4	0.119	0.364
Sum	4.000	4.000

In parallel analysis the eigenvalues in successive pairs from the actual and the randomly ordered data are compared. Factors are retained whenever the eigenvalue for the actual data for a given factor exceeds the eigenvalue for the related factor for the randomly ordered scores. These eigenvalues for the Table 3.2 data and the Table 3.3 correlation matrices are presented in Table 3.4.

For this example, because 2.776 is greater than 1.756, the first factor from the actual data would be extracted. Because 0.828 is less than 1.082, the second factor would not be extracted.

Brian O'Connor (2000) has written several SPSS and SAS syntax files to execute variations on these analyses. The programs can also be downloaded from **http://flash.lakeheadu.ca/~boconno2/nfactors.html**

FACTOR EXTRACTION METHOD

There are numerous statistical theories that can be used to compute factor pattern coefficients. Probably the most frequently used EFA extraction method, perhaps because it is the default analysis in most statistical packages, is called *principal components analysis* (see Russell, 2002).

Principal components analysis assumes that the scores on measured variables have perfect reliability. Of course, scores are never perfectly reliable (see Thompson, 2003). Furthermore, the analysis attempts to reproduce the variance or information in the *sample* data, rather than the population. Of course, if the sample is reasonably representative of the population data, the sample factors should tend to match the population factors.

Principal components analysis, which actually was the analysis described in chapter 2, uses ones on the diagonal of the correlation matrix. This seems reasonable to the extent that scores on a given measured variable should correlate perfectly with the same scores on the same variable.

However, if scores on a variable are not perfectly reliable, then correlations of scores on a measured variable with itself will not be perfect (i.e., +1.0) if measurement error is taken into account. Logically, it might be reasonable to alter the diagonal entries of the correlation matrix to take measurement error into account.

It was noted previously that communality coefficients are lower-bound (or conservative) estimates of score reliability. Another common factor extraction method, *principal axes factor analysis,* uses the communality coefficients to replace the ones on the diagonal of the correlation matrix.

Principal axes factor analysis often starts with a principal components analysis. Then the communality coefficients from this analysis are substituted respectively for the ones on the diagonal of the original correlation matrix. It is important to emphasize that no off-diagonal data are altered in this process. Then a new set of factors and their corresponding communality coefficients are estimated.

This process is successively repeated until the communality estimates stabilize. This process is called *iteration*. Iteration is a statistical process of estimating a set of results, and then successively tweaking results until two consecutive sets of estimates are sufficiently close that the iteration is deemed to have converged. Iteration is used in various ways throughout multivariate statistics, including in other aspects of EFA, such as factor rotation, which is discussed in the next section.

The principal axes communality iteration process does not always converge. Estimates may sequentially bounce up, bounce down, bounce up, and so forth. If this happens, often the sample size is too small for the number of measured variables and the model being estimated.

Always carefully check whether iteration has converged. Some computer programs print only innocuous-sounding warnings about failure to converge, and then still print full results. So the unwary analyst may not notice a convergence problem. Programs also have built-in maximum numbers of iterations (e.g., 25). If a solution does not converge, override the default number of iterations. If a solution does not converge in 100 or 125 iterations, usually the solution will not converge at any number of iterations. Either a different factor extraction method must be selected or sample size must be increased.

Principal components analysis does *not* have to be the first step of the principal axes factor analysis. Any reliability estimates that the researcher deems reasonable can be used as initial estimates. For example, if the measured variables are test scores, the reliability coefficients reported either in test manuals or in a prior "reliability generalization" (RG) analysis (Vacha-Haase, 1998) can be used as initial estimates placed on the diagonal of the correlation matrix.

Another strategy for creating initial estimates of communality coefficients is available in most statistical packages. The multiple R^2 of a given measured variable with all the other measured variables is used as the initial estimate for that variable's h^2 (e.g., $h_1^2 = R^2_{PER1 \times PER2,PER5,PER6}$, $h_2^2 = R^2_{PER2 \times PER1,PER5,PER6}$).

In theory, what one uses as the initial communality coefficients in this iteration process will affect how many iterations must be performed (with principal components invoking ones on the diagonal of the correlation matrix obviously taking the most iterations) but not what are the final estimates. If this is so, given that the only downside of numerous iterations is wasting a little electricity, it doesn't matter too much which initial estimation procedure is used.

Other factor extraction methods include alpha factor analysis, maximum likelihood factor analysis, image factor analysis, and canonical factor analysis. *Alpha factor analysis* focuses on creating factors with maximum reliability. *Maximum likelihood analysis* focuses on creating factors that reproduce the correlation or covariance matrix in the population, versus in the sample. *Image factor analysis* focuses on creating factors of the latent variables that exclude or minimize "unique factors" consisting of essentially only one measured variable. *Canonical factor analysis* seeks to identify factors that are maximally related to the measured variables.

FACTOR ROTATION

Factor rotation involves moving the factor axes measuring the locations of the measured variables in the factor space so that the nature of the underlying constructs becomes more obvious to the researcher. Factor rotation can be performed either graphically, guided only by the researcher's visual judgment, or empirically using one of the dozens of different algorithms that have been developed over the years. Analytical rotation has some advantages over graphical rotation, because the procedure is automated, and because any two researchers analyzing the same data will arrive at the same factor structure when using the same empirical rotation method.

TABLE 3.5
Correlation and Squared Correlation Coefficients for 6 Variables and 100 Participants

Variable	PER1	PER2	PER3	PER5	PER6	PER7
PER1	1.00	*72.3%*	*60.8%*	16.0%	10.2%	23.0%
PER2	.85	1.00	*51.8%*	20.3%	10.2%	25.0%
PER3	.78	.72	1.00	26.0%	16.0%	25.0%
PER5	.40	.45	.51	1.00	*43.6%*	*64.0%*
PER6	.32	.32	.40	.66	1.00	*50.4%*
PER7	.48	.50	.50	.80	.71	1.00

Note. The (unsquared) correlation coefficients are presented in the bottom triangle. The squared correlation coefficients are presented as percentages in the top triangle; values above 43% are italicized.

Rotation is not possible when only one factor is extracted. But in virtually all cases involving two or more factors, rotation is usually essential to interpretation.

The first 100 scores on six LibQUAL+™ variables (PER1, PER2, PER3, PER5, PER6, and PER7) presented in Appendix A can be used to demonstrate why rotation is so essential in EFA. Table 3.5 presents the bivariate r and r^2 values associated with these data.

An examination of the r^2 values in Table 3.5 (remember that r values must be squared before their magnitudes can be compared in an interval metric) suggests (a) the existence of two factors, consisting respectively of PER1, PER2, and PER3, and of PER5, PER6, and PER7, because all pairs of the variables in these two sets have r^2 values greater than 43%, and (b) the two factors probably are not highly correlated, because the r^2 values between measured variables across the two sets are all less than 27%. The unrotated pattern/structure coefficients and the eigenvalues presented in Table 3.6 confirm the expectation of two factors. The two factors reproduce 84.2% (3.81 + 1.23 = 5.05 / 6 = .842) of the variance in the correlation matrix involving the six measured variables.

However, the unrotated factor pattern/structure coefficients presented in Table 3.6 do *not* match our expectations regarding the composition of the factors. For example, as reported in the table, all six variables are correlated .70 or more with Factor I! Yet the Figure 3.2 plot of the factor space, based on the pattern/structure coefficients reported in Table 3.6, *do* match our expectations regarding factor structure. This suggests that the patterns in our unrotated results are exactly what we expect, but that the patterns cannot be readily discerned from statistical results in the form of unrotated results.

The idea of rotation is that the locations of the axes in the factor space are completely arbitrary. We can move (i.e., rotate) these axes to any

TABLE 3.6
Unrotated Principal Components Pattern/Structure Coefficients

Variable	Unrotated factors		Squared structure		
	I	II	I	II	h²
PER1	0.81	−0.50	65.9%	25.3%	91.2%
PER2	0.81	−0.46	65.6%	20.9%	86.5%
PER3	0.82	−0.35	67.9%	12.1%	80.0%
PER5	0.80	0.43	63.6%	18.8%	82.4%
PER6	0.70	0.54	49.2%	29.7%	78.9%
PER7	0.83	0.41	69.3%	16.4%	85.7%
λ			3.81	1.23	

Note. Pattern/structure coefficients ≥ |.35| are italicized. For the principal components model the eigenvalues equal the sums of the squared unrotated pattern/structure coefficients for a given factor (e.g., .659 + .656 + .679 + .636 + .492 + .693 = 3.81).

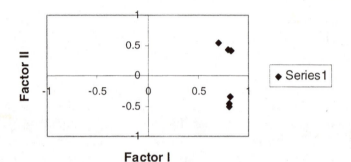

Figure 3.2. Table 3.6 variables presented in factor space before rotation.

location, and in any manner we wish. However, we may not move the location of any variable in the factor space.

Persons first learning of rotation are often squeamish about the ethics of this procedure. Certainly ethics have an important bearing on the exercise of statistical choice. If we have several mice in a drug experiment, and Elwood in the experimental group is not responding well to the new intervention, it is unethical to take Elwood out for a little walk at 3:00 a.m. that results in Elwood's "accidental" demise. That is unethical scientific conduct (and cruelty to Elwood). But factor rotation is *not* in any respect unethical (as long as we only move factor axes and not measured variables).

Table 3.7 presents an example of an empirical rotation of the Table 3.6 pattern/structure coefficient matrix. Note that (a) the communality coefficients for each of the variables remain unaltered by rotation and (b) the total variance reproduced in the factors as a set (84.2% = [2.59 + 2.46 = 5.05 / 6 = .842]) remains unaltered.

TABLE 3.7
Rotated Principal Components Pattern/Structure Coefficients

Variable	Rotated factors		Squared structure		
	I	II	I	II	h^2
PER1	0.94	0.19	87.5%	3.7%	91.2%
PER2	0.90	0.23	81.4%	5.1%	86.5%
PER3	0.84	0.31	70.1%	9.9%	80.0%
PER5	0.28	0.86	7.9%	74.5%	82.4%
PER6	0.13	0.88	1.8%	77.1%	78.9%
PER7	0.32	0.87	10.6%	75.2%	85.7%
Trace			2.59	2.46	

Note. Pattern/structure coefficients ≥ |.35| are italicized. For the principal components model the trace variance reproduced by the rotated factors equals the sums of the squared rotated pattern/structure coefficients for a given factor (e.g., .875 + .814 + .701 + .079 + .018 + .106 = 2.59).

However, it is true that rotation tends to redistribute the location of variance within the solution. In the unrotated solution the first factor reproduces the bulk of the observed variance (3.81 / 6 = 63.5%). In the rotated solution the two rotated factors reproduce 43.2% (2.59 / 6) and 41.0% (2.46 / 6) of the observed variance. The rotation has almost equalized the location of the information within the solution. This tends to happen to the extent that rotations are performed in pursuit of "simple structure."

Thurstone (1935, p. 156, 1947, p. 335) proposed the basic concept of simple structure and the associated criteria for evaluating simple structure. Think about what set of factor pattern coefficients would maximize our ability to interpret the nature of the latent constructs underlying scores on measured variables.

The unrotated results presented in Table 3.6 were virtually impossible to interpret because all of the measured variables were very highly correlated with both factors. A more interpretation-friendly solution would strike a balance between having enough measured variables correlated with a factor that some measured variables reveal the underlying nature of a factor, but not so many that the factors cannot be distinguished from each other or are so complex that their nature cannot easily be apprehended.

A simple structure would involve from a column perspective roughly an equal number of measured variables being appreciably correlated with each factor (e.g., 6 / 2 = 3), but from a row perspective most or all measured variables being appreciably correlated with only one factor. These criteria are met very well for the rotated factor solution presented in Table 3.7.

Empirical (i.e., nongraphical) rotation methods perform the rotation by applying an algorithm to derive a transformation matrix ($\mathbf{T}_{F \times F}$) by which the unrotated pattern coefficient matrix ($\mathbf{P}_{V \times F}$) is postmultiplied to yield the rotated factors. There are dozens of formulas for deriving $\mathbf{T}_{F \times F}$ matrices, several of which are available in various statistical packages. These computer programs will print the transformation matrices on request. The more a transformation matrix deviates from being an identity matrix, the more the unrotated versus the rotated factor pattern coefficients will be different.

Orthogonal Rotation

Figure 3.2 and Table 3.7 involve a case in which the original 90-degree angle between the two factors was maintained in the rotation process. This means that the rotated factors will remain "orthogonal" or uncorrelated. Factors are always orthogonal upon extraction, and remain uncorrelated if orthogonal rotation is used.

The correlation coefficient actually is a function of the cosine of the angle between either measured or composite variables. When two variables have been standardized to have a length or variance of 1.0, as factors are, the cosine of the angle between two factors is the correlation of the factors (see Gorsuch, 1983, chap. 4).

The most common orthogonal rotation method, and indeed the most common rotation of *any* kind, is the *varimax* rotation method developed by Kaiser (1958). This is the default rotation in most statistical packages. Varimax tends to focus on maximizing the differences between the squared pattern/structure coefficients on a factor (i.e., focuses on a column perspective). In my experience, in about 85% of exploratory factor analyses varimax will yield simple structure.

Another orthogonal rotation method is *quartimax*, which tends to focus on simple structure primarily from the perspective of rows (i.e., for variables across the factors). Quartimax is most appropriate when the first factor is expected to be a large "G" or general factor that is saturated by disproportionately many of the variables. *Equamax* is a compromise between varimax and quartimax methods.

Oblique Rotation

In some cases simple structure cannot be obtained using orthogonal rotation. Typically this is indicated by variables having pattern/structure coefficients that are large in absolute value on two or more factors (which is sometimes called multivocal vs. univocal). Such factors are difficult to interpret. The researcher must then turn to oblique rotation in the pursuit of simple structure.

However, as noted in chapter 2, when factors are rotated orthogonally, the pattern coefficient and the structure coefficient matrices, which always have the same rank, also will contain exactly the same numbers. This occurs because when factors are uncorrelated the interfactor correlation matrix

($\mathbf{R}_{F \times F}$) is an identity matrix, and given that $\mathbf{P}_{V \times F} \mathbf{R}_{F \times F} = \mathbf{S}_{V \times F}$, if $\mathbf{R}_{F \times F} = \mathbf{I}_{F \times F}$, then necessarily $\mathbf{P}_{V \times F} = \mathbf{S}_{V \times F}$.

But once factors are rotated obliquely, the factors are correlated, and $\mathbf{P}_{V \times F} \neq \mathbf{S}_{V \times F}$. Correlations among the factors that deviate further from zero lead to greater differences between the pattern and the structure coefficients. And once factors are rotated obliquely, the pattern coefficients are no longer correlation coefficients. On obliquely rotated factors, any or even all of the pattern coefficients can be outside the range –1 to +1.

When orthogonal rotation is performed in computer packages, $\mathbf{R}_{F \times F}$ will not be printed, because the interfactor correlation is by definition known to be the identify matrix. And because $\mathbf{P}_{V \times F}$ equals $\mathbf{S}_{V \times F}$, only one factor matrix will be printed, which may be somewhat ubiquitously labeled something like "varimax-rotated factor matrix." When an oblique rotation is performed, all three matrices ($\mathbf{P}_{V \times F}$, $\mathbf{R}_{F \times F}$, and $\mathbf{S}_{V \times F}$) will always be printed, so that the results can be correctly interpreted by consulting both $\mathbf{P}_{V \times F}$ and $\mathbf{S}_{V \times F}$.

When oblique rotation is necessary, *promax* rotation (Hendrickson & White, 1964) is almost always a good choice. Promax is actually conducted as a series of rotations. The first rotation is an orthogonal rotation, such as varimax. These pattern/structure coefficients are then raised to some exponential power, k, which is sometimes called a pivot power. If an even-numbered pivot power (e.g., $k = 4$) is used, the signs of the resulting coefficients are restored after the exponential operation is performed.

Raising varimax pattern/structure coefficients to an exponential power makes all the resulting coefficients closer to zero. But the effect is differential across original values that are larger in magnitude versus smaller in magnitude. For example, $.9^4$ is .656, whereas $.2^4$ is .002. The resulting coefficients tend to have simple structure.

Next, a "Procrustean" rotation is invoked. Procrustean rotation can also be useful as a rotation method on its own merits and not as part of promax rotation, as when a researcher wants to rotate factors to best fit a theoretical structure (Thompson & Pitts, 1982), or when conducting an EFA meta-analysis (Thompson, 1989). Indeed, some test distributions have been developed for statistically evaluating the fit of actual and theoretically expected structures, or of actual structures from different studies (Thompson, 1992b).

Procrustean rotation derives its name from the Greek myth about an innkeeper with only one bed. If guests were too short for the bed, he stretched the guests on a rack. If guests were too tall for the bed, he cut off parts of their legs.

Procrustean rotation finds the best fit of an actual matrix to some target matrix (e.g., a hypothetically expected pattern matrix, the factor solution for the same variables in a prior study). In the promax use of

Procrustean rotation, the varimax pattern/structure coefficients each raised to the pivot power become the target. The resulting Procrustean rotation is the promax solution.

The researcher can control the degree of correlation among the factors ($\mathbf{R}_{F \times F}$) by changing the pivot power. The higher the exponent, the greater will be the degree of correlation among the factors.

Another oblique rotation method is *oblimin*. Oblimin invokes a value called *delta* to control the degree of correlation among the rotated factors. Delta values of zero yield factors that are more highly correlated, whereas large negative values produce factors that are more uncorrelated. However, in cases where oblique solutions are required, promax is usually a very good choice.

FACTOR SCORES

As noted in chapter 1, sometimes factor analysis is used not to identify and interpret factors, but to obtain a more parsimonious set of composite scores (i.e., factor scores) that are then used in subsequent analyses (e.g., regression) instead of the measured variable scores. There are four methods of obtaining factor scores (i.e., composite variable scores for each person on each factor) that are briefly summarized here. Chapter 4 includes more detailed comparisons of these methods.

The most commonly used method of obtaining factor scores ($\mathbf{F}_{N \times V}$) is called the *regression* method. As a first step, the measured variables are each converted into z scores with means of zero and *SD*s of 1.0. Then the following algorithm is applied:

$$\mathbf{F}_{N \times F} = \mathbf{Z}_{N \times V}\, \mathbf{R}_{V \times V}^{-1}\, \mathbf{P}_{V \times F}.$$

The rightmost portion of the formula can be reexpressed as:

$$\mathbf{W}_{V \times F} = \mathbf{R}_{V \times V}^{-1}\, \mathbf{P}_{V \times F}.$$

This is the "factor score matrix output" by SPSS.

Second, Bartlett (1937) proposed methods for estimating factor scores intended to minimize the influence of the unique factors consisting of single measured variables not usually extracted in the analysis. Third, Anderson and Rubin (1956) proposed methods similar to Bartlett's methods, but with an added constraint that the resulting factor scores must be orthogonal.

All of these factor score estimation procedures yield factor scores that are themselves z scores, if the extraction method is principal components, or the estimation method is Anderson–Rubin. This does not create difficulties if

TABLE 3.8
$R_{4 \times 4}^{-1}$, $P_{4 \times 2}$, and $W_{4 \times 2}$ Matrices for Four Table 3.2 Variables

Variable	$R_{4 \times 4}^{-1}$				$P_{4 \times 2}$		$W_{4 \times 2}$	
	PER1	PER2	PER5	PER6	I	II	I	II
PER1	4.711	−3.664	0.049	−1.028	.90	.33	.55080	−.11583
PER2	−3.664	4.151	−0.453	0.543	.94	.22	.62861	−.22316
PER5	0.049	−0.453	2.288	−1.507	.26	.89	−.17846	.60668
PER6	−1.028	0.543	−1.507	2.430	.26	.90	−.18121	.61055

Note. The bivariate correlation matrix for these data was presented in the bottom triangle of Table 3.3.

the subsequent analyses make comparisons of subgroups (e.g., test whether boys and girls have different factor scores on Factor I). However, if there is interest in comparing factor scores across factors *within the data set as a whole*, then these factor scores are not useful. For the full sample all factor scores have means of zero, so no differences across factors can ever be detected for the group as a whole.

Fourth, Thompson (1993a) presented a method of computing factor scores that can be used when the focus is on comparing factor score means across factors for the data set as a whole. These are called "standardized, noncentered factor scores."

Table 3.8 presents the inverted correlation matrix ($R_{4 \times 4}^{-1}$) and the varimax-rotated pattern/structure coefficients ($P_{4 \times 2}$) for a principal component analysis of four of Appendix A variables (PER1, PER2, PER5, and PER6). The varimax rotation did converge, and did so in three iterations. Table 3.8 also presents the factor score coefficient weight matrix ($W_{4 \times 2} = R_{4 \times 4}^{-1} P_{4 \times 2}$) from this analysis.

Table 3.9 presents the four variables in both z score and standardized noncentered form for the subset of 25 participants. The weight matrix is then applied to the standardized noncentered scores, just as would be the case with regular regression estimates of factor scores, except that the regular regression estimates apply the weights to the four measured variables instead in their z score form. The SPSS syntax commands to create the standardized noncentered scores and the two sets of standardized noncentered factor scores can be typed into the syntax file as:

```
descriptives per1 to per6/save .
compute cPER1=zPER1 + 7.08 .
compute cPER2=zPER2 + 7.04 .
compute cPER5=zPER5 + 5.32 .
compute cPER6=zPER6 + 5.40 .
print formats zper1 to cper6 (F7.2) .
list variables=id zper1 to cper6/cases=99 .
```

TABLE 3.9
z Scores and Standardized Noncentered Scores for the
Four Measured Variables

ID	z scores				Standardized noncentered			
	ZPER1	ZPER2	ZPER5	ZPER6	CPER1	CPER2	CPER5	CPER6
1	.48	−.02	−1.33	−1.59	7.56	7.02	3.99	3.81
2	−1.08	−.02	−.75	−.19	6.00	7.02	4.57	5.21
3	−.56	−1.11	−.18	−1.12	6.52	5.93	5.14	4.28
4	−1.08	−1.11	−.75	−.65	6.00	5.93	4.57	4.75
5	−1.08	−1.11	−.18	−.65	6.00	5.93	5.14	4.75
6	−.04	−.02	.96	1.21	7.04	7.02	6.28	6.61
7	.48	.52	.39	.75	7.56	7.56	5.71	6.15
8	−.04	−.02	−.18	−1.12	7.04	7.02	5.14	4.28
9	−3.14	−2.20	−2.47	−2.06	3.94	4.84	2.85	3.34
10	.99	1.07	−.18	.75	8.07	8.11	5.14	6.15
11	.99	1.07	.96	1.68	8.07	8.11	6.28	7.08
12	−1.59	−2.20	.39	−.19	5.49	4.84	5.71	5.21
13	−.56	−.57	.96	.75	6.52	6.47	6.28	6.15
14	−.04	−.02	−1.90	−1.12	7.04	7.02	3.42	4.28
15	.99	1.07	.96	−.19	8.07	8.11	6.28	5.21
16	.99	1.07	−.18	−.19	8.07	8.11	5.14	5.21
17	.48	.52	.96	.28	7.56	7.56	6.28	5.68
18	.48	−1.11	−.75	1.21	7.56	5.93	4.57	6.61
19	.99	1.07	.39	.75	8.07	8.11	5.71	6.15
20	.48	−.02	−.18	.28	7.56	7.02	5.14	5.68
21	.99	1.07	2.10	1.68	8.07	8.11	7.42	7.08
22	−.04	−.02	.96	.28	7.04	7.02	6.28	5.68
23	−.04	.52	−.18	.28	7.04	7.56	5.14	5.68
24	.48	1.07	−.18	−1.12	7.56	8.11	5.14	4.28
25	.48	.52	.39	.28	7.56	7.56	5.71	5.68
Mean	0.00	0.00	0.00	0.00	7.08	7.04	5.32	5.40
SD	1.00	1.00	1.00	1.00	1.00	1.00	1.00	1.00

Note. The means of the standardized noncentered scores are the same as the means of these variables in their original form, as reported in Table 3.2.

```
descriptives variables=cper1 to cper6 .
compute btscr1= (.55080 * cper1) + (.62861 * cper2)
   + (−.17846 * cper5) + (−.18121 * cper6) .
compute btscr2=(−.11583 * cper1) + (−.22316 * cper2)
   + (.60668 * cper5) + (.61055 * cper6) .
print formats fscr1 fscr2 btscr1 btscr2(F8.3) .
```

Table 3.10 presents both the regular regression factor scores and the standardized noncentered factor scores for these data. Both sets of factor scores are compared here only for heuristic purposes. In actual research the score estimation method would be selected on the basis of the research questions and the related preferences for different factor score properties.

TABLE 3.10

Conventional Regression Factor Scores ($\mathbf{F}_{25 \times 2}$) and Standardized
Noncentered Factor Scores for the EFA of the Four Measured Variables

	Regression		Standardized noncentered	
ID	FSCR1	FSCR2	BTSCR1	BTSCR2
1	.773	−1.824	7.170	2.309
2	−.437	−.442	5.960	3.691
3	−.770	−.483	5.627	3.650
4	−1.037	−.485	5.360	3.649
5	−1.139	−.138	5.258	3.996
6	−.428	1.334	5.969	5.467
7	.386	.520	6.783	4.654
8	.199	−.786	6.596	3.348
9	−2.300	−1.898	4.097	2.235
10	1.115	−.008	7.512	4.126
11	.741	1.256	7.138	5.390
12	−2.295	.797	4.102	4.930
13	−.970	1.230	5.427	5.363
14	.505	−1.826	6.903	2.307
15	1.080	.115	7.477	4.249
16	1.284	−.578	7.681	3.555
17	.368	.582	6.765	4.715
18	−.522	.476	5.875	4.610
19	1.013	.339	7.410	4.472
20	.230	.010	6.627	4.143
21	.537	1.950	6.934	6.083
22	−.259	.763	6.138	4.897
23	.288	−.052	6.685	4.082
24	1.169	−1.089	7.566	3.045
25	.470	.235	6.867	4.369
Mean	0.000	0.000	6.397	4.133
SD	1.000	1.000	1.000	1.000

The standardized noncentered factor scores are perfectly correlated with their regular regression factor score counterparts (i.e., $r_{FSCR1 \times BTSCR1} = 1.0$ and $r_{FSCR2 \times BTSCR2} = 1.0$), and the correlations across the two factor scores also match across the two score types (i.e., $r_{FSCR1 \times FSCR2} = r_{BTSCR1 \times BTSCR2}$, which here is 0.0). For the Table 3.10 results, a comparison of the means on the standardized noncentered factor scores suggests that the 25 participants on the average rated their libraries higher on Factor I (6.397) than on Factor II (4.133).

MAJOR CONCEPTS

The sequence of EFA involves numerous decisions (i.e., what association matrix to analyze, how many factors to extract, factor extraction method, factor rotation method, factor score estimation method). For each of the steps of analysis, there are myriad possible decisions or ways of informing judgment. Thus, across all the steps and all the choices for each step, the possible number of combinations of analytic choices is *huge*!

It is also important not to limit the analysis at any stage to a single analytic choice. For example, in deciding how many factors to extract, researchers do well to consult several procedures (e.g., scree plot, parallel analysis) and then evaluate whether various strategies corroborate each other. The most confidence can be vested in solutions for which factors are stable or invariant over various analytic choices.

The purpose of EFA is to isolate factors that have simple structure or, in other words, are interpretable. Any thoughtful analytic choices that yield clear factors are justified, including analyzing the data in lots of different ways to "let the data speak."

In practice, factor rotation is necessary. The only exception is when a single factor is extracted, in which case no rotation is possible. Nor in such a case is rotation needed, because presumably every variable will saturate the single factor, and the factor is whatever explains all the variables being highly associated with each other. Usually when multiple factors are extracted, reasonable simple structure is realized with varimax rotation. And if these factors are interpretable, the correct rotation method has typically been identified. However, even when an orthogonal rotation yields simple structure, when the purpose of the analysis is to obtain factor scores for use in subsequent statistical analyses, if the researcher believes that the latent constructs have noteworthy correlations, an oblique rotation (e.g., promax) may be suitable.

4

COMPONENTS VERSUS
FACTORS CONTROVERSIES

In the previous chapter it was noted that there are myriad methods with which to extract factors, some of which are available in commonly used statistics packages. It is important to understand the extent to which extraction choices may affect factor interpretations, and, especially, to understand *why* and *when* differences may arise across methods for a given data set.

For reasons that will be explained momentarily, it might be argued that

> when the number of variables is [at least] moderately large, for example, greater than 30, and the analysis contains virtually no variables expected to have low communalities (e.g., .4), practically any of the exploratory [extraction] procedures ... will lead to the same interpretations. (Gorsuch, 1983, p. 123)

Similarly, Cliff (1987) suggested that "the choice of common factors or components methods often makes virtually no difference to the conclusions of a study" (p. 349). These dynamics will be illustrated here with data for different variables and participants from the Appendix A scores.

Table 4.1 compares three of the more elegant extraction methods that, unlike principal components analysis, each invokes iteration as part of the extraction process: alpha, maximum likelihood, and image factor analysis. As an inspection of the tabled results suggests, for these data the three extraction methods yield reasonably comparable factor structures.

TABLE 4.1

Varimax-Rotated Pattern/Structure Coefficient Matrices for Three Factor
Extraction Methods (*n* = 100 Faculty)

Variable	Alpha			ML			Image		
	I	II	III	I	II	III	I	II	III
PER1	.20	*.90*	.18	.20	*.91*	.16	.21	*.84*	.19
PER2	.12	*.82*	.24	.11	*.86*	.19	.12	*.82*	.20
PER3	.23	*.62*	.22	.26	*.64*	.12	.24	*.61*	.19
PER4	.16	*.86*	.26	.19	*.84*	.20	.18	*.80*	.24
PER5	*.89*	.13	.22	*.90*	.13	.18	*.86*	.14	.20
PER6	*.87*	.14	.22	*.87*	.15	.19	*.84*	.16	.20
PER7	*.92*	.18	.20	*.92*	.20	.09	*.86*	.19	.18
PER8	*.72*	.31	.18	*.72*	.32	.16	*.71*	.32	.16
PER9	.34	.18	*.44*	.39	.22	.32	.37	.21	.35
PER10	.21	.23	*.72*	.20	.23	*.95*	.25	.31	*.45*
PER11	.17	.34	*.56*	.23	.37	.40	.21	.36	*.42*

Note. Pattern/structure coefficients greater than |.4| are italicized. ML = maximum likelihood.

The largest difference occurs on the third factor, which involves only 2 or 3 of the 11 measured variables. And the results from the image factor analysis differ somewhat more from the first two sets of results than these two sets differ from each other. This is partly because image factor analysis focuses on minimizing the influence of factors involving single variables not reflected via covariations in the "images" of the other measured variables.

However, the differences across extraction methods tend to decrease as there are more measured variables. For example, regarding image analysis, as more measured variables are added there is an increasing likelihood that most measured variables will be correlated with another measured variable or a combination of the other measured variables, so the influences of "specific" factors consisting of a single measured variable, and not representing true latent variables, will tend to become less as there are more measured variables. Bear in mind that the present small heuristic analysis involving only 11 measured variables is unrealistic. Researchers would rarely (if ever) factor analyze so few measured variables.

MEASUREMENT ERROR VERSUS SAMPLING ERROR

As noted in chapter 3, principal axes factor analysis is the same as principal components analysis, except that in principal axes factoring *only* the diagonal of the correlation matrix is modified by removing the 1.0's originally always residing there and iteratively estimating the true commu-

TABLE 4.2
Communality Coefficients (h^2) Through Four Iterations
(n = 100 Graduate Students)

Variable	R^2	Iteration			
		1	2	3	4
PER1	.83714	.89374	.92230	.93880	.94904
PER2	.74596	.76327	.76468	.76277	.76061
PER3	.67464	.69098	.69167	.68996	.68843
PER4	.65751	.67103	.67066	.66857	.66692
PER5	.75740	.79961	.81378	.81939	.82197
PER6	.55652	.58434	.58885	.58870	.58798
PER7	.73583	.77671	.78818	.79155	.79257
PER8	.65091	.67977	.68493	.68521	.68479

nality coefficients. This is done to take into consideration the fact that all scores have some measurement error (i.e., scores are never perfectly reliable).

Table 4.2 illustrates the iteration process for a principal axes factor analysis involving the first eight measured variables presented in Appendix A, and limiting the analysis to the 100 graduate students (i.e., the first 100 scores in Appendix A). In this particular analysis, before iteration was begun, R^2 values were substituted on the diagonal of the correlation matrix. These results involve the R^2 between a given measured variable and the seven remaining measured variables in the analysis.

For example, as reported in Table 4.2, the R^2 for predicting the 100 scores on the measured variable, PER1, with the scores on the other seven measured variables in this analysis (i.e., PER2, PER3, . . . PER8) was .83714. The R^2 for predicting the 100 scores on the measured variable, PER2, with the scores on the other seven measured variables in this analysis (i.e., PER1, PER3, . . . PER8) was .74596.

Using the R^2 values in this manner is sensible, although other initial estimates of communality may also be plausible, because the reliabilities of the scores do impact the multiple regression effect size (Thompson, 2003). If a particular variable was perfectly unreliable (i.e., measured "nothing," or only random fluctuations in scores), then the R^2 would theoretically be zero.

Table 4.2 then presents the h^2 estimates for each of four successive steps of iteration. These can be obtained in SPSS by invoking the CRITERIA subcommand:

```
factor variables=per1 to per12/
  analysis=per1 to per8/
  criteria=factors(2) iterate(1)
```

```
econverge(.003)/print=all/
extraction=paf/rotation=varimax/ .
```

Notice that the estimates tend to change less and less across successive iterations.

Because scores are never perfectly reliable (Thompson, 2003), it seems reasonable to extract factors in a manner that takes into account inherent imperfections in measurement. However, there are actually several types of error that impact any statistical analysis.

Measurement error is the variance in scores that is deemed unreliable. This area-world proportion can be estimated as 1.0 minus a reliability coefficient.

But *sampling error* also affects all statistical analyses, whenever the researcher has only data from a subset of the population (i.e., a sample) and not the complete population. Sampling error is the unique, idiosyncratic variance in a given sample that does not exist (a) in the population, which has no idiosyncratic variance due to a sampling process, or (b) in other samples, which do have sampling error, but sampling error variances that are unique each to themselves. Just as every person is different, every sample has its own personality or sampling error.

The more statistical estimates we make in any analysis for a given data set, the greater is the likelihood that we are capitalizing on the unique sampling error variance in a given sample of scores. Because sampling error variance is unique to a given sample, this implies also that our results will be less and less generalizable either to the population or to future samples as we make more and more estimates. Clearly, some reticence in invoking endless estimation seems appropriate!

When we are iteratively estimating communality coefficients, to take into account measurement error variance, we are also allowing sampling error variance to affect our results. There is a trade-off between taking into account these two error sources.

Because communality estimates change the most in the first few iterations, and sampling error variance is more influential as we conduct more iterations, it may be reasonable to limit these iterations to two, or three, or four (Snook & Gorsuch, 1989). Of course, there are no magic answers regarding how to strike this balance.

The comment by Huberty and Morris (1988, p. 573) is worth remembering: "As in all of statistical inference, subjective judgment cannot be avoided. Neither can reasonableness!" The important thing is to be aware of these dynamics, regardless of what balance is struck. Of course, when sample size is very large, and the number of measured variables is smaller, sampling error is smaller (Thompson, 2002), so more iterations may be reasonable in the presence of large samples.

By far the two most commonly used estimation methods are principal components and principal axes factors analyses. Principal components analysis is the default method in commonly used statistical packages, and is used with considerable frequency whenever exploratory factor analyses (EFAs) are performed (Fabrigar et al., 1999; Russell, 2002). The *only* difference between these two methods is that 1.0's are used on the diagonal of the correlation matrix in principal components analysis. In principal axes factoring, the communality estimates on the diagonal of the correlation matrix are iteratively estimated until the iterations converge (i.e., remain fairly stable across a contiguous pair of estimates).

It would be a major understatement to say that "analysts differ quite heatedly over the utility of principal components as compared to common or principal factor analysis" (Thompson & Daniel, 1996, p. 201). As Cliff (1987) noted,

> Some authorities insist that component analysis is the only suitable approach, and that the common factors methods just superimpose a lot of extraneous mumbo jumbo, dealing with fundamentally unmeasurable things, the common factors. Feelings are, if anything, even stronger on the other side. . . . Some even insist that the term "factor analysis" must not even be used when a components analysis is performed. (p. 349)

Indeed, an entire special issue on this controversy was published in the journal, *Multivariate Behavioral Research* (Mulaik, 1992). And Gorsuch (2003) recently noted that "in following the dialogue for the past 30 years, the only major change seems to be that the heat of the debate has increased" (p. 155).

A series of comparisons of principal components versus principal axes factors for the Appendix A data makes this discussion concrete. Table 4.3 presents a principal components versus a principal axes factor analysis of

TABLE 4.3
Components Versus Principal Axes Factors for Four Measured Variables
(*n* = 100 Graduate Students)

Variable	Varimax-rotated components			Varimax-rotated factors		
	I	II	h^2	I	II	h^2
PER1	.94	.19	92.3%	.88	.22	83.0%
PER2	.94	.22	92.7%	.90	.25	87.2%
PER5	.28	.86	82.1%	.27	.79	70.4%
PER6	.13	.92	85.3%	.16	.78	62.9%

Note. Pattern/structure coefficients greater than |.4| are italicized. Given the small number of variables, the convergence criterion for principal axes factor extraction was raised from the default of .001 to .003.

TABLE 4.4

Varimax-Rotated Components Versus Factors for Eight Measured Variables With Factors Estimated with R^2 Starting Values and 1.0 Starting Values (n = 100 Graduate Students)

Variable	Components			Principal-axes factors[a]			Principal-axes factors[b]		
	I	II	h^2	I	II	h^2	I	II	h^2
PER1	.94	.19	91.5%	.96333	.18636	96.3%	.96362	.18625	96.3%
PER2	.88	.23	82.8%	.82979	.26108	75.7%	.82971	.26112	75.7%
PER3	.82	.32	77.2%	.75473	.34148	68.6%	.75466	.34151	68.6%
PER4	.82	.29	76.1%	.74967	.32049	66.5%	.74959	.32054	66.5%
PER5	.27	.88	84.0%	.26440	.86862	82.4%	.26447	.86841	82.4%
PER6	.15	.84	73.0%	.18993	.74214	58.7%	.18985	.74235	58.7%
PER7	.33	.85	82.6%	.32802	.82784	79.3%	.32802	.82782	79.3%
PER8	.27	.83	76.7%	.28089	.77795	68.4%	.28089	.77796	68.4%

Note. Pattern/structure coefficients greater than |.4| are italicized.
[a]Communality estimation was begun using R^2 starting values.
[b]Communality estimation was begun using 1.0 starting values.

four measured variables. Although the patterns of the results (two measured variables saturate each of the two latent variables) are similar, two differences appear.

First, the pattern/structure coefficients tend to be smaller for the principal axes factors. Second, the communality coefficients tend to be smaller for the principal axes factors. These two outcomes are a function of changing *only* the diagonal of the correlation matrix from 1.0s to smaller positive numbers within the principal axes analyses.

Table 4.4 presents a principal components analysis versus two principal axes factor analyses of eight measured variables. With regard to the principal axes factor analyses, the first analysis again began with h^2 estimation using R^2 values of each measured variable with the remaining seven measured variables, and estimation continued until convergence was achieved in seven iterations. The second principal axes factor analysis began h^2 estimation using 1.0 values (i.e., the unaltered correlation matrix was used as a starting point), and estimation continued until convergence was achieved in eight iterations. Clearly the two sets of results are comparable, even for this example with only eight measured variables, which is why these results had to be reported to more than two decimal values.

This is the general pattern in regard to selecting starting points for iteration. Theoretically, roughly the same values will be realized regardless of the initial estimates. However, using less reasonable initial estimates (i.e., 1.0's vs. R^2 or reliability coefficients) will require more iterations. On a modern microcomputer, the additional computations from more iteration will usually go unnoticed.

The interpretations of the Table 4.4 components versus factors would be the same, even though the h^2 values tend to be larger for the component analysis. And the two sets of results are more comparable for the Table 4.4 comparisons involving eight measured variables than for the Table 4.3 comparisons involving four measured variables. Why?

Two factors affect the equivalence of the principal components and the principal axes factors for a given data set. First, if the reliabilities of the scores being analyzed are quite high (i.e., approach 1.0), then the h^2 values will tend to each be roughly 1.0, which is the same h^2 value used for each variable in the principal components analysis. This might occur, for example, if the researcher was analyzing measured variables that were each scores on tests each consisting of a large number of items.

Second, there is also a less obvious dynamic at work. The *only* difference between principal components and principal axes factor analyses involves changing the diagonal of the correlation matrix when principal axes factor analyses are conducted. But the impact of altering only the diagonal of the correlation matrix differs depending on how many measured variables are being analyzed!

As Snook and Gorsuch (1989, p. 149) explained, "As the number of variables decreases, the ratio of diagonal to off-diagonal elements also decreases, and therefore the value of the communality has an increasing effect on the analysis." For example, with four measured variables, there are 16 (4×4) entries in the correlation matrix, of which four are on the diagonal ($4 / 16 = 25.0\%$). When there are eight measured variables, there are 64 (8×8) entries in the correlation matrix, of which eight are on the diagonal ($8 / 64 = 12.5\%$).

When there are 100 measured variables, there are 10,000 entries in the correlation matrix, of which 100 are on the diagonal ($100 / 10,000 = 1.0\%$). And it is more common to factor analyze 100 than 4 measured variables! As there are more measured variables, even if their reliabilities are modest, principal components and principal axes factors tend to be increasingly similar (Ogasawara, 2000). And principal components have some additional desirable properties, as demonstrated in chapter 5, that makes these analyses attractive, at least to some researchers.

MAJOR CONCEPTS

There are numerous EFA factor extraction methods (e.g., alpha, canonical, image, maximum likelihood). However, most frequently seen in published research are principal components and principal axes methods. Principal components is the default method in widely used statistical packages.

Principal axes methods acknowledge measurement error by iteratively approximating the communality estimates on the diagonal of the correlation matrix. When score reliability is high for the measured variables, there will be little difference between the factors extracted using these two methods.

In any case, the differences in the two methods will also be smaller as the number of factored variables increases. This is because iterative estimation of commonalities changes *only* the diagonal entries of the correlation matrix, and the diagonal of the matrix is progressively a smaller constituent of the correlation matrix as more measured variables are analyzed (e.g., 4 diagonal entries with 4 measured variables is 4 out of 16 [25.0%] matrix entries; 8 diagonal entries with 8 measured variables is 8 out of 64 [12.5%] matrix entries). And in practice we usually factor fairly large numbers of measured variables.

There is also a trade-off in considering measurement error variances. Each sample data set also contains sampling error variance. The more steps we iterate to take into account measurement error, the more statistical estimates we are making, and the more opportunity we are creating for sampling error to influence our results.

5

COMPARISONS OF SOME
EXTRACTION, ROTATION,
AND SCORE METHODS

The exploratory factor analysis (EFA) statistics called structure coefficients were first introduced in chapter 2, and the statistics called communality coefficients were first introduced in chapter 3. Both of these statistics merit further discussion. This chapter further elaborates the concepts underlying structure and communality (h^2) coefficients by invoking factor score statistics. First, however, some comparisons of factor score estimation methods may be helpful.

FACTOR SCORES IN DIFFERENT EXTRACTION METHODS

As noted in chapter 4, the two most commonly used EFA approaches are principal components and principal axes factor analyses. And, as noted in chapter 3, there are four different methods with which to estimate factor scores. Here three factor score methods (regression, Bartlett, and Anderson–Rubin) are compared to facilitate deeper understanding of the factor score estimation choices and to elaborate additional differences in the two most common extraction methods.

In real research, factor scores are typically only estimated when the researcher elects to use these scores in further substantive analyses (e.g., a multivariate analysis of variance comparing mean differences on three factors across men and women). In these applications the researcher will use only one factor score estimation method. Here several methods are compared simultaneously only for heuristic purposes.

Factor scores can be derived by invoking SPSS COMPUTE statements that apply the factor score coefficient matrix ($\mathbf{W}_{V \times F}$) to the measured variables in either z score or standardized, noncentered form. This has the heuristic value of making this computation process concrete and explicit. However, in actual research it is both easier and more accurate to obtain the factor scores by using the SPSS SAVE subcommand within the FACTOR procedure.

For example, to obtain regression factor scores from a principal components analysis involving 11 measured variables (PER1 through PER11), the syntax would be:

```
factor variables=per1 to per12/
   analysis=per1 to per11/criteria=factors(3)/
   extraction=pc/rotation=varimax/
   print=all/save=reg(all reg_pc)/ .
```

The execution of this commands yields three factor scores for each participant from a principal components analysis that are labeled REG_PC1, REG_PC2, and REG_PC3.

Factor scores from a principal axes factor analysis estimated using the Bartlett method would be produced with the syntax:

```
factor variables=per1 to per12/
   analysis=per1 to per11/criteria=factors(3)/
   extraction=paf/rotation=varimax/
   print=all/save=bart(all bar_paf)/ .
```

Factor scores were computed for all 100 faculty using the first 11 measured variables presented in Appendix A for all six combinations of two extraction methods (principal components vs. principal axes factor analysis) and three factor score estimation methods (regression, Bartlett, and Anderson–Rubin). Table 5.1 presents the three sets of principal components factor scores for the first 5 and last 5 participants in the data set. Notice that *when principal components methods are used, all three factor score methods yield identical factor scores* for a given participant (e.g., for participant #101, REG_PC1 = BAR_PC1 = AR_PC1 = +.35).

Table 5.2 presents the three sets of principal axes analysis factor scores for the first 5 and last 5 participants in the data set. Notice that when

TABLE 5.1
Selected Factors Scores for Varimax-Rotated Principal Components Analyses of 11 Measured Variables (PER1 Through PER11) (*n* = 100 Faculty)

ID	Regression			Bartlett			Anderson–Rubin		
	REG_PC1	REG_PC2	REG_PC3	BAR_PC1	BAR_PC2	BAR_PC3	AR_PC1	AR_PC2	AR_PC3
101	.35	-.51	.18	.35	-.51	.18	.35	-.51	.18
102	-.58	-.35	.04	-.58	-.35	.04	-.58	-.35	.04
103	.44	-1.82	-.67	.44	-1.82	-.67	.44	-1.82	-.67
104	-.16	1.33	.18	-.16	1.33	.18	-.16	1.33	.18
105	-.77	.93	.23	-.77	.93	.23	-.77	.93	.23
...									
195	.56	-1.58	1.35	.56	-1.58	1.35	.56	-1.58	1.35
196	.10	-1.47	-1.38	.10	-1.47	-1.38	.10	-1.47	-1.38
197	.98	1.24	-.79	.98	1.24	-.79	.98	1.24	-.79
198	-.87	-1.26	-.91	-.87	-1.26	-.91	-.87	-1.26	-.91
199	1.37	-2.45	1.56	1.37	-2.45	1.56	1.37	-2.45	1.56
200	-1.78	-1.57	-.83	-1.78	-1.57	-.83	-1.78	-1.57	-.83

TABLE 5.2
Selected Factors Scores for Varimax-Rotated Principal Axes Factor Analyses of 11 Measured Variables (PER1 Through PER11) (n = 100 Faculty)

ID	Regression			Bartlett			Anderson–Rubin		
	RG_PAF1	RG_PAF2	RG_PAF3	BR_PAF1	BR_PAF2	BR_PAF3	AR_PAF1	AR_PAF2	AR_PAF3
101	.31	-.53	.16	.31	-.57	.27	.31	-.53	.20
102	-.76	-.18	-.03	-.86	-.07	.06	-.83	-.07	.00
103	.27	-1.90	-.36	.29	-2.03	-.28	.27	-1.95	-.31
104	-.09	1.12	.26	-.13	1.17	.30	-.11	1.13	.29
105	-.78	1.02	.00	-.81	1.18	-.14	-.79	1.11	-.09
⋮									
195	.64	-1.40	.92	.55	-1.76	1.68	.60	-1.59	1.26
196	-.05	-1.21	-1.23	.08	-1.08	-1.87	-.01	-1.13	-1.52
197	.82	1.00	-.33	.94	1.18	-.75	.88	1.09	-.49
198	-.88	-1.15	-.90	-.87	-1.17	-1.07	-.88	-1.17	-.94
199	1.35	-1.81	.79	1.37	-2.26	1.50	1.38	-2.08	1.13
200	-1.69	-2.06	-.57	-1.76	-2.20	-.42	-1.72	-2.12	-.51

principal axes factor analyses methods are used, all three factor score methods yield *different* factor scores for a given participant (e.g., for participant #102, RG_PAF1 = −.76; BR_PAF1 = −.86; AR_PAF1 = −.83).

Table 5.3 presents the Pearson correlation coefficients among all pairwise combinations of the 18 factor scores for the 100 participants in this analysis. Recall that all six analyses invoked rotation to the varimax criterion, which means that the three factors in all six analyses were perfectly uncorrelated.

As the correlation coefficients in Table 5.3 indicate, the correlations of the factor scores from the principal components analyses exactly matched the correlations (here all zero) of the factors themselves. This *always* (and only always) happens in principal components analysis. The exact matching of factor score correlations with factor correlations is another appealing feature of principal components analyses.

STRUCTURE COEFFICIENTS IN A COMPONENTS CONTEXT

As noted in chapter 2, correlations between measured variables and composite variables are always called *structure coefficients* (r_S) throughout the general linear model (GLM; Courville & Thompson, 2001). Table 5.4 presents the Pearson correlation coefficients for the 11 measured variables with the three factor scores from the principal components analysis. These were obtained with the SPSS syntax:

```
correlations variables=per1 to per11 with
  reg_pc1 to reg_pc3 .
```

Note that the Table 5.4 structure coefficients literally computed as bivariate correlation coefficients exactly match the pattern/structure coefficients produced by the SPSS principal components analysis of the same 11 measured variables. These varimax-rotated pattern/structure coefficients are presented in Table 5.5.

COMMUNALITY COEFFICIENTS AS R^2 VALUES

In chapter 2 it was noted that communality coefficients indicate the proportion of variance of a measured variable that the factors as a set can reproduce. Conversely, h^2 can be viewed as characterizing how much of the variance of a measured variable was useful in delineating the extracted factors.

TABLE 5.3
Factor Score Correlations (*n* = 100 Faculty)

| | Principal components | | | | | | | | |
| | Regression | | | Bartlett | | | Anderson–Rubin | | |
Variable	REG_PC1	REG_PC2	REG_PC3	BAR_PC1	BAR_PC2	BAR_PC3	AR_PC1	AR_PC2	AR_PC3
REG_PC1	1.0000								
REG_PC2	.0000	1.0000							
REG_PC3	.0000	.0000	1.0000						
BAR_PC1	1.0000	.0000	.0000|	1.0000					
BAR_PC2	.0000	1.0000	.0000|	.0000	1.0000				
BAR_PC3	.0000	.0000	1.0000|	.0000	.0000	1.0000			
AR_PC1	1.0000	.0000	.0000|	1.0000	.0000	.0000|	1.0000		
AR_PC2	.0000	1.0000	.0000|	.0000	1.0000	.0000|	.0000	1.0000	
AR_PC3	.0000	.0000	1.0000|	.0000	.0000	1.0000|	.0000	.0000	1.0000
RG_PAF1	.9903	−.0017	.0596|	.9903	−.0017	.0596|	.9903	−.0017	.0596|
RG_PAF2	.0054	.9779	.0543|	.0054	.9779	.0543|	.0054	.9779	.0543|
RG_PAF3	.0485	.1007	.9571|	.0485	.1007	.9571|	.0485	.1007	.9571|
BR_PAF1	.9904	−.0064	−.0314|	.9904	−.0064	−.0314|	.9904	−.0064	−.0314|
BR_PAF2	.0061	.9748	−.0577|	.0061	.9748	−.0577|	.0061	.9748	−.0577|
BR_PAF3	−.0476	−.0116	.9501|	−.0476	−.0116	.9501|	−.0476	−.0116	.9501|
AR_PAF1	.9917	−.0050	.0208|	.9917	−.0050	.0208|	.9917	−.0050	.0208|
AR_PAF2	.0044	.9793	.0035|	.0044	.9793	.0035|	.0044	.9793	.0035|
AR_PAF3	−.0023	.0383	.9535|	−.0023	.0383	.9535|	−.0023	.0383	.9535|

One heuristic to make these concepts concrete is to conduct a multiple regression analysis in which the measured variables in turn are predicted by the factor scores from the principal components analysis. These regression analyses can be executed using the SPSS syntax:

```
regression variables=reg_pc1 to reg_pc3
  per1 to per11/dependent=per1 /
  enter reg_pc1 to reg_pc3 .
regression variables=reg_pc1 to reg_pc3
  per1 to per11/dependent=per2 /
  enter reg_pc1 to reg_pc3 .
. . .
regression variables=reg_pc1 to reg_pc3
  per1 to per11/dependent=per11 /
  enter reg_pc1 to reg_pc3 .
```

TABLE 5.3 (Continued)
Factor Score Correlations (n = 100 Faculty)

	Principal axes							
Regression			Bartlett			Anderson–Rubin		
RG_PAF1	RG_PAF2	RG_PAF3	BR_PAF1	BR_PAF2	BR_PAF3	AR_PAF1	AR_PAF2	AR_PAF3
1.0000								
.0050	1.0000							
.0997	.1136	1.0000						
.9945	.0000	.0000	1.0000					
.0000	.9923	.0000	.0058	1.0000				
.0000	.0000	.9869	−.1000	−.1151	1.0000			
.9985	.0021	.0545	.9985	.0016	−.0454	1.0000		
.0013	.9976	.0639	.0007	.9979	−.0511	.0000	1.0000	
.0455	.0512	.9949	−.0547	−.0642	.9977	.0000	.0000	1.0000

Figure 5.1 presents the SPSS output from predicting the first two measured variables, respectively, by the three factor scores from the principal components analysis of the 11 measured variables with the 100 faculty members as participants. Note from Figure 5.1 that the R^2 for the measured variable PER1 predicted by REG_PC1, REG_PC2, and REG_PC3 is .87990. This matches the h^2 value for this measured variable presented in Table 5.5. Because the three factor scores are uncorrelated, the beta weights for the three factor scores are also the correlation coefficients between the predictors and the outcome variable (Courville & Thompson, 2001), and thus also equal the structure coefficients for the three measured variables presented in Table 5.5.

Similarly, the R^2 for the measured variable PER2 predicted by REG_PC1, REG_PC2, and REG_PC3 is .82204. This matches the h^2 value for this measured variable presented in Table 5.5. Hopefully, these illustrations bring home the concepts underlying factor scores, factor structure coefficients, and communality coefficients.

TABLE 5.4
Pearson *r* Coefficients of 11 Measured Variables With Factor Scores
From Principal Components Analysis (*n* = 100 Faculty)

Measured variable	Factor score		
	REG_PC1	REG_PC2	REG_PC3
PER1	.2028	.8966	.1868
PER2	.1083	.8740	.2154
PER3	.2305	.7318	.1620
PER4	.1636	.8729	.2410
PER5	.9081	.1265	.2172
PER6	.8965	.1347	.2164
PER7	.9083	.1779	.1987
PER8	.7934	.3239	.1448
PER9	.3366	.1053	.6677
PER10	.1819	.2294	.7919
PER11	.1190	.3394	.7197

TABLE 5.5
Varimax-Rotated Pattern/Structure Coefficients From Principal
Components Analysis of the 11 Measured Variables (*n* = 100 Faculty)

Variable	Component			h^2
	I	II	III	
PER1	.20284	*.89659*	.18675	87.990%
PER2	.10828	*.87403*	.21537	82.204%
PER3	.23051	*.73176*	.16198	61.485%
PER4	.16361	*.87285*	.24099	84.671%
PER5	*.90807*	.12646	.21722	88.777%
PER6	*.89654*	.13470	.21639	86.875%
PER7	*.90832*	.17788	.19870	89.617%
PER8	*.79337*	.32388	.14477	75.528%
PER9	.33664	.10530	*.66769*	57.022%
PER10	.18185	.22943	*.79192*	71.285%
PER11	.11897	.33942	*.71967*	64.728%

Note. Pattern/structure coefficients greater than |.4| are italicized.

MAJOR CONCEPTS

There are numerous ways to calculate scores for each person on each factor (i.e., factor scores). These composite or latent variable scores can be computed and then used in subsequent analyses (e.g., as variables in a descriptive discriminant analysis, as predictors in a regression analysis), just as measured variables can be used in these analyses. When principal components analyses are conducted, regression, Bartlett, and Anderson–Rubin factor scores for a given person on a given factor will all be the same, and in other extraction methods will differ.

```
Dependent Variable. .          PER1 1 Willingness to help users
Method:     Enter             REG_PC1 REG_PC2 REG_PC3
Multiple R                .93803
R Square                  .87990
------------------------------------ Variables in the Equation -------------------------------------

Variable                  B          SE B      Beta              T        Sig T
REG_PC1              .277091      .048319    .202837        5.735       .0000
REG_PC2             1.224812      .048319    .896592       25.349       .0000
REG_PC3              .255119      .048319    .186753        5.280       .0000
(Constant)          7.650000      .048077                 159.121       .0000

End Block Number      1     All requested variables entered.

Dependent Variable. .          PER2 1 Giving users individual att
Method:     Enter             REG_PC1 REG_PC2 REG_PC3

Multiple R                .90666
R Square                  .82204
------------------------------------ Variables in the Equation -------------------------------------

Variable                  B          SE B      Beta              T        Sig T
REG_PC1              .167676      .066672    .108283        2.515       .0136
REG_PC2             1.353442      .066672    .874029       20.300       .0000
REG_PC3              .333508      .066672    .215374        5.002       .0000
(Constant)          7.310000      .066338                 110.194       .0000
```

Figure 5.1. Multiple regression predicting measured variables using the three factor scores (*n* = 100 faculty).

Factor scores can also be used for heuristic purposes, to better grasp the nature of both structure coefficients and communality coefficients. Throughout the GLM, a structure coefficient, score-world statistic, can be computed (literally) as the bivariate product–moment correlation of a measured variable (e.g., the scores of n people on a factored variable) with a latent variable (e.g., the principal components factor scores of n people). And a communality coefficient, an area-world statistic, in the orthogonal principal components case can be computed as the R^2 between the factor scores of n people with the scores of n people on the measured variables.

6

OBLIQUE ROTATIONS AND HIGHER-ORDER FACTORS

In chapter 3 the notion of factor rotation was introduced. It was noted that rotation is always possible in exploratory factor analysis (EFA) whenever there is more than one factor, and that rotation is almost always necessary in these cases to obtain simple structure. One class of methods involves orthogonal rotation and leaves the factors still perfectly uncorrelated. A second class of methods involves oblique rotation and allows the unrotated factors (that are initially always uncorrelated) to become correlated. Oblique rotations present some interpretation challenges that merit further discussion.

Table 5.5 (see chap. 5) presented the varimax-rotated principal components from an analysis of the first 11 measured variables presented in Appendix A for data provided by the 100 faculty members. This solution had reasonably simple structure. A few measured variables were not quite as clear in regard to what factors they reflected. For example, PER11 had a pattern/structure coefficient on Factor III of .72, but on Factor II had a pattern/structure coefficient of .34.

However, it is important to remember that all correlation coefficients, including structure coefficients, must be squared before their magnitudes can be compared. Thus, because $.72^2$ is 51.8%, and $.34^2$ is 11.6%, the larger of these two structure coefficients was actually 4.5 times larger than the smaller value.

TABLE 6.1
Promax-Rotated Components for 11 Measured Variables (n = 100 Faculty)

Variable	Pattern			Structure			h^2
	I	II	III	I	II	III	
PER1	.04	*.94*	−.03	.40	.94	.46	88.0%
PER2	−.07	*.92*	.03	.31	.91	.46	82.2%
PER3	.10	*.75*	−.03	.39	.78	.40	61.5%
PER4	−.01	*.90*	.05	.37	.92	.50	84.7%
PER5	*.95*	−.06	.04	.94	.34	.46	88.8%
PER6	*.93*	−.05	.04	.93	.34	.45	86.9%
PER7	*.94*	.00	.00	.95	.38	.45	89.7%
PER8	*.81*	.20	−.07	.85	.49	.41	75.5%
PER9	.20	−.12	*.70*	.48	.31	.73	57.0%
PER10	−.03	.01	*.85*	.38	.43	.84	71.2%
PER11	−.10	.17	*.75*	.33	.51	.79	64.7%

Note. Pattern coefficients greater than |.4| are italicized.

Although the varimax-rotated solution had simple structure, for heuristic purposes the solution was also rotated to the promax criterion. Because $S_{V \times F}$ equals $P_{V \times F}(R_{F \times F})$, and $R_{F \times F}$ now will not equal $I_{F \times F}$, the structure and the pattern coefficient matrices will no longer be equal. Indeed, because the pattern coefficients are no longer correlation coefficients, some, many, or all of the pattern coefficients may be less than −1 or greater than +1.

Table 6.1 presents the pattern and the structure matrices from this analysis. Table 6.2 presents the factor correlation matrix from the analysis.

In chapter 2 the notion of the communality coefficient was first introduced. In actuality, the example involved a varimax-rotated principal components solution. In an orthogonal rotation the h^2 can be computed by summing the squared structure coefficients for a variable across the factors. This yields the total variance of the measured variable common to the factors as a set. The computation is simplified because the factors are uncorrelated, and so there is zero overlap between the common variances of a given measured variable on any two factors.

But when factors are correlated, the h^2 is derived by first computing the products of a given measured variable's pattern coefficient on each

TABLE 6.2
Factor Correlation Matrix ($R_{3 \times 3}$) for the Table 6.1 Factor Pattern Matrix

Factor	I	II	III
I	1.000		
II	.398	1.000	
III	.473	.502	1.000

factor with the corresponding structure coefficient on each factor, and then summing these products. For example, for the measured variable PER1 in the Table 6.1 solution, h^2 for this measured variable is computed as:

$$[0.040 \times 0.399] + [0.936 \times 0.937] + [-0.029 \times 0.460] =$$
$$0.016 + 0.877 + -0.013 = 87.96\%.$$

It should be noted that the algorithm for a given measured variable, $\Sigma P_K S_K$ across the k factors, is a more general formula for h^2. When factors are uncorrelated, $\Sigma P_K S_K$ across the k factors can be reexpressed as ΣP_K^2, because in this case the pattern coefficients are also structure coefficients. In other words, for principal components and principal axes analyses the algorithm $\Sigma P_K S_K$ across the k factors always works, whether or not the factors are correlated.

MAXIMIZING SAMPLE FIT VERSUS REPLICABILITY

In the present comparison the Table 5.5 varimax-rotated solution and the Table 6.1 promax-rotated solution would generally lead to the same interpretations regarding the makeup of the three underlying constructs. Where the solutions differ is in the representations of the correlations among the latent constructs.

The first component in Table 5.5 would be interpreted as measuring the "Library as Place." This component was saturated with perceptions of libraries including respectively the items "7. A contemplative environment," "5. A haven for quiet and solitude," "6. A meditative place," and "8. Space that facilitates quiet study."

The second component might be labeled "Service Affect." This component was saturated respectively with the items "1. Willingness to help users," "2. Giving users individual attention," "4. Employees who are consistently courteous," and "3. Employees who deal with users in a caring fashion."

The third component might be labeled "Information Access." This component was saturated respectively with the items "10. Complete runs of journal titles," "11. Interdisciplinary library needs being addressed," and "9. Comprehensive print collections."

But for these data the promax-rotated results reported in Table 6.1 would lead to the same factor labels as those produced in the varimax-rotated solution. The pattern coefficients in Table 6.1, like the pattern/structure coefficients presented in Table 5.5, had simple structure.

The structure coefficients in Table 6.1 reflect a dynamic involving the components being fairly highly correlated, as reported in Table 6.2. Most of the structure coefficients in Table 6.1 are quite noteworthy for all variables on all components. This is merely indicative of the factors having correlations of .398, .473, and .502.

Of course, the components are sufficiently discrete that it still seems reasonable to recognize the existence of three latent variables, because even the two most highly correlated components only have 25.2% common variance ($.502^2 = .252$). But for a case such as this, how would one select a factor analytic solution? Three related issues seem particularly relevant.

Analytic Purpose

In some factor analytic applications the researcher is primarily interested in interpreting the factors. In other applications, factor scores are generated for use in subsequent statistical analyses. As the present example suggests, when factor interpretation is the focus, rotation choice may be less critical as long as a given analysis yields a reasonable simple structure.

But when composite scores are going to be created for use in subsequent analyses, and when the oblique factors are highly correlated, some preference might be given to the oblique solution, so that the correlations of the latent variables will be honored in these subsequent substantive analyses. However, there is one other approach to estimating factor scores, sometimes called *unit weighting*, that may finesse these concerns (see Gorsuch, 1983, pp. 268–276).

As noted in chapter 4, every sample contains some idiosyncratic and nonreproducible variance as a function of sampling error. The more parameters we estimate, the more potential there is for our results to capitalize excessively on unique features of the sample, with the consequence that results will generalize less well to new samples. One way to capitalize less on idiosyncratic features of the sample is to compute factor score estimates without using the factor score weight matrix ($\mathbf{W}_{V \times F}$).

One approach simply identifies those measured variables that definitively measure a single factor. Then composite scores on each factor are computed merely by summing the scores on the relevant measured variables, by averaging these scores, by averaging the z scores of only the relevant variables, or by doing whatever else the researcher deems thoughtful and reasonable. For example, across both the varimax-rotated and the promax-rotated results for the present example, the same measured variables would be used to obtain unit-weighted composite scores on the three constructs.

Parsimony

Centuries ago a philosopher named William of Occam articulated the concept we recognize today as the law of parsimony. Thus, this precept is sometimes called "Occam's razor." Basically, the law of parsimony says that when two explanations fit a set of facts roughly equally well, all things equal, the simpler explanation is more likely to be true. We tend to prefer parsimonious explanations because things that are true are also more likely to replicate in future research, and researchers usually do not want to be embarrassed by making discoveries that no one else can replicate.

In an EFA context, orthogonal solutions are more parsimonious than their oblique counterparts because fewer parameters are estimated when an orthogonal rotation is implemented. This is because when $\mathbf{R}_{F \times F}$ equals $\mathbf{I}_{F \times F}$, then $\mathbf{P}_{V \times F}$ equals $\mathbf{S}_{V \times F}$.

In any statistical analysis, including EFA, what we are trying to do is fit a model to our data. The more parameters we estimate, the greater is the likelihood that our model will fit the sample data. For example, if n is 2, for any two measured variables, r^2 is always +1. For a regression analysis, when the number of predictor variables equals $n - 1$, the degrees of freedom error is zero, and the resulting R^2 is always +1, regardless of what the predictor variables are.

The same situation applies across different analyses. For example, in predictive discriminant analysis (PDA), when less parsimonious equations called quadratic versus linear rules are used, the prediction hit rate tends to be higher in the original sample, but the prediction efficacy of the same prediction equation across other samples tends to be lower than in the original sample. Thus, Huberty (1994, p. 64) noted a general preference for the more parsimonious PDA equations, even though the fit of the model to sample data tends to be better for the less parsimonious solution. Elsewhere he noted that "the linear rule is suggested because of presumed greater stability [across samples] than that for the quadratic rule, in general" (p. 260).

Interpretation Difficulty

It must also be said that factor interpretation itself is more complex when factors are correlated. This is because more parameter estimates must be simultaneously interpreted. Specifically, throughout the general linear model (GLM), whenever weights are not structure coefficients, *both* standardized weights and structure coefficients must be examined to arrive at correct interpretations.

This is true with reference to analyses including multiple regression (cf. Courville & Thompson, 2001; Thompson & Borrello, 1985), canonical correlation analysis (Meredith, 1964; Thompson, 1984), and confirmatory factor analysis (cf. Graham et al., 2003; Thompson, 1997). This is also true in EFA. As Gorsuch (1983) emphasized, "*proper interpretation of a set of factors can probably only occur if at least **S** and **P** are both examined*" (p. 208).

A variable does not contribute to a factor only if the variable's pattern and structure coefficients are both zero. If a structure coefficient is near zero, but the pattern coefficient is not, the measured variable contributes to the solution indirectly by modifying the relationship of other measured variables with the factor. Across the GLM this is called a *suppressor effect* (see Courville & Thompson, 2001; Horst, 1966).

For the Table 6.1 results, the analysis does not present overwhelming difficulties, because the structure is so clear. The three factors ("Library as Place," "Service Affect," and "Information Access") were highly correlated but still sufficiently discrete to be recognized as the constructs described by the pattern coefficients. But the structure coefficients in this case clearly indicate that all 11 variables are correlated with all three constructs. Such findings suggest the presence of overarching superordinate constructs, or higher-order factors.

HIGHER-ORDER FACTORS

When *variables* are correlated (i.e., $\mathbf{R}_{V \times V} \neq \mathbf{I}_{V \times V}$), then factors can be computed as weighted aggregates of the measured variables. By the same token, when *factors* are correlated (i.e., $\mathbf{R}_{F \times F} \neq \mathbf{I}_{F \times F}$), then factors can be extracted from this matrix as well.

The factors extracted from intervariable correlations (or other statistics measuring associations) are called *first-order* factors. The factors then extracted from the interfactor correlations among the first-order factors are called *second-order* factors. If the second-order factors are correlated, then *third-order* factors can be extracted. However, as Kerlinger (1984) noted, "while ordinary factor analysis is probably well understood, second-order factor analysis, a vitally important part of the analysis, seems not to be widely known and understood" (p. xivv).

It is argued here that these higher-order factors *should* be extracted whenever factors are correlated. As Gorsuch (1983) emphasized:

> Rotating obliquely in factor analysis implies that the factors do overlap and that there are, therefore, broader areas of generalizability than just a primary factor. Implicit in all oblique rotations are higher-order factors. It is recommended that these [always] be extracted and examined so that the investigator may gain the fullest possible understanding of the data. (p. 255)

Too few researchers reporting correlated first-order factors conduct these needed higher-order analyses.

The different levels of factor analysis give different perspectives on the data. Thompson (1990b) offered the analogy of hiking in the mountains versus flying over them. When you are hiking you can see rocks and trees and fish in streams. When you fly over the mountains, you can see the patterns made by the mountains as a range. The perspectives complement

TABLE 6.3
Second-Order Factor Pattern/Structure Coefficient Matrix

First-order factor	A	h^2
FACT_I	.773	59.8%
FACT_II	.792	62.7%
FACT_III	.832	69.2%

each other, and each is needed to see patterns at a given level of specificity versus generality.

The literature includes various examples of higher-order factor analyses (cf. Borrello & Thompson, 1990; Cook, Heath, & Thompson, 2001; Cook & Thompson, 2000; Thompson & Borrello, 1986). Here both the implementation of a higher-order analysis using SPSS and the interpretation of such an analysis are illustrated.

Example #1

For the Table 6.2 interfactor correlation matrix ($\mathbf{R}_{3 \times 3}$) the second-order factors can be easily computed in SPSS by analyzing the correlations among the first-order principal components scores. The syntax is:

```
subtitle '5 n=100 Faculty ^^^ First-order ^^^^^^'.
execute .
temporary .
select if (ranktype eq 3) .
factor variables=per1 to per12/
  analysis=per1 to per11/criteria=factors(3)/
  extraction=pc/rotation=promax/
  print=all/save=reg(all reg_pc)/ .
subtitle '6 Second-order $$$$$$$$$$$$$$$$$$$$$$$$$$' .
execute .
factor variables=reg_pc1 to reg_pc3/
  print=all .
```

This works *only* for principal components analysis, because only in principal components analysis will the factor correlations and the factor score correlations match exactly. The second-order factor derived in this manner is presented in Table 6.3.

Higher-Order Interpretation Strategies

The interpretation of higher-order factors poses some special difficulties. Factors are abstractions of measured variables. Second-order factors, then, are abstractions of abstractions even more removed from the measured variables. Somehow we would like to interpret the second-order factors in terms of the measured variables, rather than as a manifestation of the factors of the measured variables.

As Gorsuch (1983) suggested,

> To avoid basing interpretations upon interpretations of interpretations, the relationships of the original variables to each level of the higher-order factors are determined. . . . Interpreting from the variables should improve the theoretical understanding of the data and produce a better identification of each higher-order factor. (pp. 245–246)

There are three strategies for doing so.

First, Gorsuch (1983) suggested that the second-order factors directly reflecting the measured variables could be evaluated by computing the product matrix of the first-order pattern matrix times the second-order pattern matrix (i.e., here $P_{11 \times 3}P_{3 \times 1} = P_{11 \times 1}$). Second, Thompson (1990b) reasoned that an orthogonally rotated product matrix might be interpreted. He reasoned that if rotation is sensible in both the first and the second order, then rotation of the product matrix to achieve simple structure might also be useful.

Of course, in the present example involving only one second-order factor, rotation is not possible. Rotation is only possible when there are two or more factors.

Third, Schmid and Leiman (1957) proposed an elegant method for expressing both the first-order and the second-order factors in terms of the measured variables, but also residualizing (removing) all variance in the first-order factors that is also present in the second-order factors. This allows the researcher to determine what, if any, variance is unique to a given level of analysis or perspective. Schmid and Leiman (1957) suggested that their Schmid-Leiman solution "not only preserves the desired interpretation characteristics of the oblique solution, but also discloses the hierarchical structuring of the variables" (p. 53).

This analysis can be implemented using SPSS matrix commands. The input into the analysis includes (a) the obliquely rotated first-order pattern matrix (here $P_{11 \times 3}$) and (b) the second-order pattern matrix (here $P_{3 \times 1}$) augmented by a symmetrical (here 3×3) matrix containing zeroes off-diagonal and the square roots of uniquenesses on the diagonal.

This augmented matrix, A, is of rank f by $s + f$, where f is the number of first-order factors and s is the number of second-order factors. In the current example, this augmented matrix is $A_{3 \times 4}$.

Recall that a communality coefficient (h^2) indicates in an area-world squared metric the percentage of variance in a measured variable that is

TABLE 6.4
Augmented Matrix $\mathbf{A}_{3 \times 4}$

SECOND	SQRT of uniquenesses		
	u_I	u_{II}	u_{III}
0.773	0.634	0.000	0.000
0.792	0.000	0.611	0.000
0.832	0.000	0.000	0.554

Note. SQRT (0.402) = 0.634.

contained in the factors as a set. The uniqueness (u^2) is an area-world statistic indicating the percentage of variance in a measured variable *not* contained in the factors as a set (e.g., $100\% - h^2 = u^2$, or $100\% - 59.8\% = 40.2\%$). Table 6.4 presents the matrix, $\mathbf{A}_{3 \times 4}$, for this analysis.

The SPSS syntax file for the present example is:

```
MATRIX .
COMPUTE P1 =
{0.040,   0.936,  -0.029 ;
-0.070,   0.917,   0.031 ;
 0.103,   0.751,  -0.026 ;
-0.014,   0.900,   0.050 ;
 0.947,  -0.061,   0.037 ;
 0.933,  -0.049,   0.037 ;
 0.944,   0.003,   0.003 ;
 0.805,   0.202,  -0.070 ;
 0.198,  -0.121,   0.700 ;
-0.026,   0.010,   0.851 ;
-0.096,   0.165,   0.753} .
COMPUTE P2 =
{0.773 ;
 0.792 ;
 0.832} .
COMPUTE Pvh = P1 * P2 .
COMPUTE A =
{0.773, 0.634, 0.000, 0.000 ;
 0.792, 0.000, 0.611, 0.000 ;
 0.832, 0.000, 0.000, 0.554} .
COMPUTE SL = P1 * A .
PRINT P1 /
  FORMAT='F8.3' /
  TITLE='First-order Promax-rotated Pattern
Matrix' /
```

```
      SPACE=4 /
      RLABELS=PER1, PER2, PER3, PER4, PER5,
      PER6, PER7, PER8, PER9, PER10, PER11 /
      CLABELS=FACT_I, FACT_II, FACT_III / .
   PRINT P2 /
      FORMAT='F8.3' /
      TITLE='Second-order Pattern Matrix' /
      SPACE=4 /
      RLABELS=FACT_I, FACT_II, FACT_III /
      CLABELS=SECOND / .
   PRINT Pvh /
      FORMAT='F8.3' /
      TITLE='Pattern Product Matrix' /
      SPACE=4 /
      RLABELS=PER1, PER2, PER3, PER4, PER5,
      PER6, PER7, PER8, PER9, PER10, PER11 /
      CLABELS=SECOND / .
   PRINT A /
      FORMAT='F8.3' /
      TITLE='The A Matrix' /
      SPACE=4 /
      RLABELS=FACT_I, FACT_II, FACT_III /
      CLABELS=SECOND U1 U2 U3 / .
   PRINT SL /
      FORMAT='F8.3' /
      TITLE='Schmid-Leiman Solution' /
      SPACE=4 /
      RLABELS=PER1, PER2, PER3, PER4, PER5,
      PER6, PER7, PER8, PER9, PER10, PER11 /
      CLABELS=SECOND, FACT_I, FACT_II, FACT_III / .
   END MATRIX .
```

Note. A FORTRAN floating-point format declaration (e.g., "F8.3") says how many columns wide a number is (e.g., 8) and how many numbers are to the right of the decimal point (e.g., 3).

Table 6.5 presents the pattern product matrix for the analysis. Table 6.6 presents the Schmid-Leiman solution. Note that the product matrix ($P_{V \times S}$) presented in Table 6.5 is the same as the first column of the Schmid-Leiman solution presented in Table 6.6. This is always the case and implies that *either* the product matrix would be interpreted or the more complete Schmid-Leiman solution, but not both.

TABLE 6.5
Pattern Product Matrix ($\mathbf{P}_{11 \times 1}$)

Variable	SECOND
PER1	.748
PER2	.698
PER3	.653
PER4	.744
PER5	.715
PER6	.713
PER7	.735
PER8	.724
PER9	.640
PER10	.696
PER11	.683

TABLE 6.6
Schmid-Leiman Solution for Example #1

Variable	SECOND	First-order		
		FACT_I	FACT_II	FACT_III
PER1	.748	.025	.572	−.016
PER2	.698	−.044	.560	.017
PER3	.653	.065	.459	−.014
PER4	.744	−.009	.550	.028
PER5	.715	.600	−.037	.020
PER6	.713	.592	−.030	.020
PER7	.735	.598	.002	.002
PER8	.724	.510	.123	−.039
PER9	.640	.126	−.074	.388
PER10	.696	−.016	.006	.471
PER11	.683	−.061	.101	.417

Example #2

The second example involves perceptions of library service quality provided by 61,316 university students and faculty using the 25 LibQUAL+[TM] items presented in Table 6.7. Four first-order factors were extracted. For heuristic purposes only, in this analysis two second-order factors were extracted. The Schmid-Leiman solution is reported in Table 6.8.

The richness of higher-order factor analysis can be communicated by interpreting these results. The two second-order factors measure "Library as Place" ("SEC_B") and all other perceptions as "SEC_A." However, the nine "Service Affect" and the six "Personal Control" items particularly dominate the first second-order factor.

TABLE 6.7
SPSS Variable Names and Variable Labels for Example #2
($N = 61,316$ Students and Faculty)

Name	Variable label
PR1	SA Willingness to help users
PR4	SA Employees who are consistently courteous
PR11	SA Dependability handling users^ service probs
PR14	SA Giving users individual attention
PR15	SA Employees deal w users in a caring fashion
PR17	SA Employees knowledge to answer user questions
PR18	SA Readiness to respond to users^ questions
PR20	SA Employees who instill confidence in users
PR24	SA Employees understand needs of their users
PR2	LP Space that facilitates quiet study
PR10	LP A haven for quiet and solitude
PR13	LP A place for reflection and creativity
PR21	LP A comfortable and inviting location
PR23	LP A contemplative environment
PR3	IA Complete runs of journal titles
PR8	IA Timely document delivery/interlibrary loan
PR9	IA Interdisc library needs being addressed
PR19	IA Convenient business hours
PR22	IA Comprehensive print collections
PR5	PC elec resources accessible from home or office
PR6	PC Modern equip me easily access the info I need
PR7	PC lib website enabling me locate info on my own
PR12	PC Easy-use tools allow find things on my own
PR16	PC info easily accessible for independent use
PR25	PC Convenient access to library collections

"SA" = Service Affect; "LP" =Library as Place; "IA" = Information Access; "PC" = Personal Control. Variable labels in SPSS are limited as to number of characters. So they frequently include abbreviations and may not contain apostrophes.

Table 6.8 also reports the sums of the squared column coefficients. These values (8.531 and 4.312) suggest that the two second-order factors contain most of the reproduced variance in the solution. There is not much variance left in the first-order factors once the second-order factors are extracted and their variance is removed from the first-order factors. The sum is fairly large (1.494) for the first first-order factor, but this is largely an artifact of 9 of the 25 items being measures of "Service Affect," which is an important element at all levels in how library users evaluate service quality in that setting.

For illustrative purposes, the pattern product matrix ($\mathbf{P}_{25 \times 2}$) will also be rotated, in this example to the equamax criterion. The necessary SPSS syntax is:

TABLE 6.8
Schmid-Leiman Solution for Example #2
(*n* = 61,316 Students and Faculty)

Variable	Second-order		First-order			
	SEC_A	SEC_B	FACT_I	FACT_II	FACT_III	FACT_IV
PR1	.655	.153	.422	−.011	.024	−.045
PR4	.624	.201	.450	−.001	−.012	−.062
PR11	.690	.260	.291	.003	.072	.041
PR14	.635	.265	.403	.009	.005	−.038
PR15	.660	.256	.469	.005	−.022	−.053
PR17	.704	.234	.366	−.006	.020	.047
PR18	.713	.216	.430	−.008	.004	.007
PR20	.610	.294	.382	.006	−.057	.054
PR24	.689	.310	.378	.004	−.029	.077
PR2	.200	.754	−.025	.124	.045	−.064
PR10	.206	.820	−.028	.132	.025	−.042
PR13	.262	.771	.028	.117	.001	−.014
PR21	.328	.705	.035	.095	−.023	.086
PR23	.260	.824	.005	.120	−.043	.067
PR3	.526	.281	−.060	−.009	.010	.419
PR8	.566	.169	.048	−.018	.072	.254
PR9	.559	.274	.033	−.005	.027	.308
PR19	.416	.324	.037	.014	−.021	.247
PR22	.512	.406	−.052	.009	−.044	.449
PR5	.628	.106	−.041	−.002	.364	−.019
PR6	.685	.239	.000	.012	.306	.033
PR7	.708	.129	−.039	−.001	.399	−.016
PR12	.736	.202	.061	.003	.306	.012
PR16	.761	.231	.103	.003	.260	.044
PR25	.625	.344	.079	.010	.068	.228
Sum of squares	8.531	4.312	1.494	0.071	0.576	0.694

```
set printback=listing .
DATA LIST records=3 /
 ROWTYPE_ (A8) FACTOR_ (F2.0)
 Pr1 Pr4 Pr11 Pr14 Pr15 Pr17 Pr18 Pr20 Pr24
 Pr2 Pr10 Pr13 Pr21 Pr23
 Pr3 Pr8 Pr9 Pr19 Pr22
 Pr5 Pr6 Pr7 Pr12 Pr16 Pr25
 (1X,9F5.3/11F5.3/5F5.3).
BEGIN DATA
FACTOR 1 .655 .624 .690 .635 .660 .704 .713 .610 .689
 .200 .206 .262 .328 .260 .526 .566 .559 .416 .512 .628
 .685 .708 .736 .761 .625
```

TABLE 6.9
Equamax-Rotated Product Matrix (n = 61,316 Students and Faculty)

Variable	Second-order factor	
	A	B
PR1	.653	.159
PR4	.622	.207
PR11	.688	.266
PR14	.632	.271
PR15	.658	.262
PR17	.702	.241
PR18	.711	.223
PR20	.607	.300
PR24	.686	.316
PR2	.193	.756
PR10	.198	.822
PR13	.255	.773
PR21	.321	.708
PR23	.252	.826
PR3	.523	.286
PR8	.564	.174
PR9	.556	.279
PR19	.413	.328
PR22	.508	.411
PR5	.627	.112
PR6	.683	.245
PR7	.707	.136
PR12	.734	.209
PR16	.759	.238
PR25	.622	.350

```
FACTOR 2 .153 .201 .260 .265 .256 .234 .216 .294 .310
 .754 .820 .771 .705 .824 .281 .169 .274 .324 .406 .106
 .239 .129 .202 .231 .344
END DATA.
FACTOR
MATRIX=IN(FAC=*) /
ANALYSIS=pr1 to pr25 /
PRINT=ROTATION /
CRITERIA=ITERATE(75) /
ROTATION=equamax .
```

The equamax-rotated pattern/structure coefficients for the analysis are reported in Table 6.9. In the present example factor interpretations would be little altered by this rotation.

Of course, the product matrix would never be obliquely rotated. That would imply the existence of higher-order factors. In conducting oblique rotations, always extract the progressively higher-order factors implied by nonzero interfactor correlations, until (a) there is only a single higher-order factor or (b) the factors at the highest level of analysis have simple structure when rotated orthogonally.

MAJOR CONCEPTS

If simple structure cannot be obtained with orthogonal rotation, reflected in many variables having pattern/structure coefficients that are large in magnitude on two or more factors, oblique rotation may be needed to derive interpretable factors. However, more parameters must be estimated in oblique rotation. As a consequence, oblique results not only (a) tend to fit sample data better but also (b) tend to yield factors that replicate less well. As a consequence, unless oblique results are substantially better than orthogonal results, orthogonal result may be preferred.

Whenever oblique rotation is performed, higher-order factors are implied and should be derived and interpreted. The Schmid-Leiman solution is particularly useful in such cases. The SPSS syntax presented in the chapter invokes SPSS MATRIX procedures and can be used as a model to perform these analyses using this widely available software.

7

SIX TWO-MODE TECHNIQUES

Psychologists often think in terms of different types of people, such as workaholics or Type A personalities. Yet the factor analyses discussed in previous chapters do *not* address questions regarding types of people! Instead, all the previous discussion implicitly involved investigations of the structures underlying variables.

The previous analyses involved a raw data matrix in which people were represented as data rows and the variables were organized as column fields within the data matrix. When data are analyzed in this manner, intervariable relationships and factors of the measured variables are the focus of the analysis.

But data matrices do *not* have to be organized in this manner. For example, data could be organized such that variables define the rows and people define the columns. When data are organized in this manner, interperson relationships are the focus of the analysis, and people factors are the result. Indeed, there are other data structures that are possible as well.

The mathematics of the analysis remain the same across these variations. The computer package need not know what data features are being correlated and analyzed!

CATTELL'S "DATA BOX"

Raymond Cattell (1966a) summarized six primary exploratory factor analysis (EFA) choices in the form of his "data box." Cattell suggested that data could involve (a) either one or more than one *variable*, (b) either one or more than one *person*, and (c) either one or more than one *occasion* of measurement. These features of the data are termed *modes*.

Some data sets might involve several variables, several people, and several occasions of measurement. It is theoretically possible to do a factor analysis that simultaneously considers all three modes. Tucker (1966) provided some guidance on performing these analyses.

However, three-mode analyses are extraordinarily complex, and commonly used software does not support these methods. Thus, only two-mode analyses are routinely seen in the literature. Within each two-mode combination (e.g., variables and people, measured at only one occasion), there are two possible organizations of the data matrix (e.g., people define the rows and variables define the columns, or variables define the rows and people define the columns). Thus, there are six two-mode techniques. Each two-mode technique was labeled by Cattell using a different capital letter (e.g., R, Q).

Six Two-Mode Techniques

R-Technique

When people define the rows of the data matrix, and columns are used to represent different variables, the analysis examines the structure underlying variables. This analysis is termed *R-technique* factor analysis. All of the previous examples in this book were R-technique examples. This is the most commonly used two-mode EFA technique, and is so common that people invoking this analysis often do not even label the analysis R-technique.

Q-Technique

When variables define the rows of the data matrix, and columns are used to represent different people, the analysis identifies people factors (rather than factors of variables). This analysis is termed *Q-technique* factor analysis. Q-technique is the second most widely used EFA method.

P-Technique

When variables are factored, and occasions constitute the rows of the data matrix, the analysis is called *P-technique* factor analysis. P-technique may involve a single-person study. However, the analysis might also involve means, or medians, summarizing for each variable and each occasion all the scores as a single number.

TABLE 7.1
Six Variations of Two-Mode Factor Analysis

Technique label	Columns defining entities to be factored	Rows defining the patterns of associations	Example application
R	Variables	Participants	Thompson and Borrello (1987)
Q	Participants	Variables	Thompson (1980b)
O	Occasions	Variables	Jones, Thompson, and Miller (1980)
P	Variables	Occasions	Cattell (1953)
T	Occasions	Participants	Frankiewicz and Thompson (1979)
S	Participants	Occasions	

O-Technique

When occasions are factored, and variables constitute the rows of the data matrix, the analysis is called *O-technique* factor analysis. Like P-technique, O-technique may involve a single-person study. Again, however, the analysis might also involve means, or medians, summarizing for each variable and each occasion all the scores of several people as a single number.

T-Technique

When occasions are factored, and participants constitute the rows of the data matrix, the analysis is called *T-technique* factor analysis. This technique presumes the use of a single variable. Of course, that variable could be a highly reliable score from a well-respected measure such as an IQ or an achievement test.

S-Technique

When participants are factored, and occasions constitute the rows of the data matrix, the analysis is called *S-technique* factor analysis. Again, this technique presumes the use of a single variable, but that variable might be from a highly regarded test such as a standardized test. I am not aware of a published study reporting an S-technique factor analysis.

Illustrative Applications

However, all the first five two-mode techniques have been reported in the literature. Table 7.1 cites some of these various applications.

As noted previously, each of these analyses can be conducted with commonly available statistical software, such as SPSS. The computer does not know (or care) what data are being analyzed. All the basic precepts discussed in previous chapters as regard to the R-technique example apply across techniques.

However, this does *not* mean that the same raw data matrix can be reorganized and subjected to two different analyses (e.g., both an R- and a Q-technique analysis). There are two reasons why raw data should not be subjected to more than one two-mode method. First, the different techniques address different questions (e.g., "What are the common factors underlying variance of the variables?" or "What are the types of people?"). The analyst should use the technique that addresses the research question of interest in a given study.

Second, in every two-mode technique, the analysis begins with the computation of a matrix of associations among all the pairs of the columns (e.g., a Pearson r matrix, a Spearman rho matrix, a covariance matrix). In each of these possibilities the stability of the association statistics being analyzed is critical to the stability and replicability of the factors themselves.

Association coefficients are more accurate when the number of rows in the raw data matrix is several times larger than the number of columns being analyzed. This allows the relationship patterns to be replicated over more rows. Because the same two-mode data cannot have more rows than columns when organized in both its two possible forms (e.g., people as rows with variables as columns, and variables as rows with people as columns), it is not feasible to subject the same single data matrix to two different two-mode techniques.

Q-TECHNIQUE

Q-technique factor analysis merits further discussion and an heuristic example, because this method is used with some frequency in psychology. Stephenson's (1953) book is the definitive seminal work on this topic. Shorter accessible treatments are provided by Kerlinger (1986, chap. 32), Carr (1992), and Thompson (2000b).

Q-technique factor analysis can be used to address three research questions:

1. How many types of people are there?
2. Which people belong to the different types?
3. Which variables were the basis for delineating the different person factors?

There are various ways that data can be collected to address these three research questions in a Q-technique factor analysis.

Q-Technique Data Collection

Most published EFA studies base the analysis on the use of the Pearson *r* matrix as the matrix of association coefficients analyzed. The factors extracted from this matrix are influenced by the properties of the statistics from which the factors originate. For example, Pearson *r* values are influenced by two issues: (a) Do the two entities being correlated order the rows of the data matrix in the same order? and (b) Do the two entities being correlated have the same distribution shapes?

For the following variable pairs, the answers to both questions is "yes," and each pair of variables yields *r* = +1.0:

X	Y	X	Y	X	Y
1	1	1	2	1	100
2	2	2	4	2	200
3	3	3	6	4	400

But for the following data, even though the columns order the rows identically, the distribution shapes of the column data differ, and therefore *r* would not reach its maximum value of +1 (even though Spearman's rho for these data would be +1):

X	Y	X	Y
1	1	1	0
2	2	2	999
3	99	3	9999

In Q-technique factor analysis it is traditional to collect the data such that the columns (i.e., the data for each person) have identically the same distribution shapes. This does allow the interperson correlation coefficients potentially to reach their minimum and maximum values (−1 and +1). However, as Thompson (2000b) observed,

> it is interesting to note that researchers using Q-technique factor analysis usually are extremely careful to make sure that person scores have exactly the same distribution shapes, so that interperson relationship statistics will not be attenuated at all by distributional differences. However, persons using R-technique factor analysis often do not pay much attention to these influences as regards intervariable relationships. These differences in conventional practices across the two techniques are striking, because the mathematics of the factor extraction process are identical in both cases! (p. 215)

Q Sort

One commonly used strategy for collecting Q sort data is the Q sort. The Q sort requires participants to sort objects into a predetermined number of categories with each category containing exactly a predetermined number of objects. The sorted objects can be anything (e.g., index cards containing statements, photographs of art). The sorting criteria can vary (e.g., how much participants agree or disagree with the statements, aesthetic valuing of the art).

Typically the fixed single distribution is developed to be symmetrical and quasi-normal in shape. An example might involve sorting index cards containing statements into categories 1 (*most agree*) through 7 (*most disagree*) using the following distribution:

Category	1	2	3	4	5	6	7
n of cards	2	4	6	9	6	4	2

In this example every column of data (i.e., every person's data) would involve exactly two 1's, exactly four 2's, and so forth. The mean of *every* column would be 4.00 ($SD = 1.58$) with a coefficient of skewness of 0.00 and a coefficient of kurtosis of –0.50.

Ranking

In a ranking strategy participants are instructed to rank order the objects (e.g., index cards with statements) using some criterion (e.g., from most agreement to most disagreement). The appeal of this strategy is that the method yields greater score variance. Because more information is being collected, this, in turn, theoretically will yield more sensitive or stable interperson correlations, and thus better person factors. The downside of the strategy is that participants may find rank ordering 60 or 100 objects to be tedious and painful.

Mediated Ranking

Thompson (1980a) suggested an alternative two-stage strategy for collecting ranked data that makes the ranking process more manageable for participants. First, participants are asked to perform a conventional Q sort. Second, the participants are then asked to rank order the objects within each category. This means that fewer objects must be simultaneously considered by the participants when performing a given ranking task.

Heuristic Example

The hypothetical data presented in Table 7.2 can be used to make concrete the discussion of Q-technique interpretation issues. The example presumes that five students and four faculty rank ordered 20 features of library service quality from 1 (*most important*) to 20 (*least important*). Table

TABLE 7.2
Q-Technique Heuristic Data

	Student					Faculty				
Item	1	2	3	4	5	1	2	3	4	Item core
1	11	16	13	11	11	20	10	7	20	SA Employees who instill confidence in users
2	16	13	16	16	16	7	7	9	1	IA Convenient business hours
3	13	11	11	13	8	1	1	1	7	LP Space that facilitates quiet study
4	8	8	7	8	13	9	9	20	9	SA Employees deal users in a caring fashion
5	12	12	1	12	12	10	20	10	10	SA Employees understand needs of their users
6	7	7	8	1	7	8	8	16	8	IA Complete runs of journal titles
7	1	1	12	9	1	16	16	8	17	SA Giving users individual attention
8	15	17	15	15	9	3	3	3	16	SA Employees who are consistently courteous
9	9	9	9	7	15	17	13	17	3	IA Comprehensive print collections
10	4	4	4	4	4	12	12	13	12	SA Employees knowledge to answer user questions
11	17	15	17	17	6	13	17	12	13	PC lib web site enabling me locate info on my own
12	6	6	3	6	17	4	4	6	4	LP A contemplative environment
13	3	3	6	3	3	6	6	4	6	LP A place for reflection and creativity
14	14	10	14	19	14	15	15	11	14	IA Timely document delivery/interlibrary loan
15	10	14	10	18	10	14	11	14	15	SA Readiness to respond to users' questions
16	20	20	18	10	19	11	14	15	11	PC Convenient access to library collections
17	19	19	20	14	20	5	5	2	5	LP A haven for quiet and solitude
18	2	5	2	2	5	2	2	5	2	LP A comfortable and inviting location
19	5	2	5	5	2	19	18	19	18	SA Willingness to help users
20	18	18	19	20	18	18	19	18	19	IA Interdisc library needs being addressed

7.3 presents the varimax-rotated principal components pattern/structure coefficients from this analysis.

The first question addressed by the analysis is "How many types of people are there?" Here the answer was two. Presumably the decision to extract two factors was guided by various evaluation criteria (e.g., eigenvalues greater than 1.0, scree test, parallel analysis) as it would be in any other EFA.

TABLE 7.3
Varimax-Rotated Pattern/Structure Coefficients

	Factor	
Person	I	II
STUD1	.96	.02
STUD2	.94	−.01
STUD3	.85	.12
STUD4	.83	.23
STUD5	.75	−.14
FAC1	.03	.93
FAC2	.07	.90
FAC3	−.09	.74
FAC4	.13	.76

Note. Values greater than |.40| are presented in italics.

TABLE 7.4
Salient Factor Scores for Factor I Sorted by Descending Levels of Agreement

Item	Factor score I	II	SPSS variable label
19	−1.37	1.66	SA Willingness to help users
18	−1.35	−1.51	LP A comfortable and inviting location
13	−1.27	−.90	LP A place for reflection and creativity
10	−1.27	.39	SA Employees knowledge to answer user questions
7	−1.12	.96	SA Giving users individual attention
16	1.32	.26	PC Convenient access to library collections
20	1.48	1.52	IA Interdisc library needs being addressed
17	1.60	−1.33	LP A haven for quiet and solitude

The second question addressed is "Which people belong to the different types?" Here the five students saturate Factor I, and the four faculty saturate Factor II. The pattern/structure matrix has simple structure. (Would that real results were always so definitive!)

The third question addressed is "Which variables were the basis for delineating the different person factors?" To address this question, factor scores are computed for each variable on both the person factors. These factor scores are in z-score form, as are all factor scores except standardized, noncentered factor scores.

The variables with factor scores less than −1.0 or greater than +1.0 are more than one standard deviation in distance from the factor score means of zero. These are the items of greatest importance or least importance primarily to the persons defining the two factors. Table 7.4 presents all the factor scores for Factor I, the student person factor, for which factor scores were greater than |1.0|. Table 7.5 presents all the factor scores for Factor II, the faculty person factor, for which factor scores were greater than |1.0|.

TABLE 7.5
Salient Factor Scores for Factor II Sorted by Descending Levels of Agreement

Item	Factor score I	II	SPSS variable label
3	.26	−1.55	LP Space that facilitates quiet study
18	−1.35	−1.51	LP A comfortable and inviting location
17	1.60	−1.33	LP A haven for quiet and solitude
12	−.55	−1.28	LP A contemplative environment
20	1.48	1.52	IA Interdisc library needs being addressed
19	−1.37	1.66	SA Willingness to help users

In the present example, lower rankings (e.g., 1, 2) indicated items ranked most important. And all the noteworthy pattern/structure coefficients in Table 7.3 are positive. This means that factor scores in Tables 7.4 and 7.5 that are smallest (i.e., biggest negative values) were items deemed most important, whereas factor scores that are largest were items deemed least important.

The factor scores for Factor I presented in Table 7.4 suggest that the students were most concerned that library staff are service oriented and that the library is a comfortable place for reflection. At the other extreme, the students were least concerned about the library being quiet and about library collections.

The factor scores for Factor II presented in Table 7.5 suggest that the faculty were most concerned about the library being quiet and least concerned about the helpfulness of library staff and the interdisciplinary needs of users being addressed. These hypothetical faculty only care about whether the library coverage in their own disciplines is adequate!

Variables with large factor scores and different signs across factors reflect what views particularly differentiate the two person factors. For example, students want library staff to be helpful (F_I = −1.37), whereas this is less important to faculty (F_{II} = +1.66). However, both students and faculty want the library to be comfortable (F_I = −1.35 and F_{II} = −1.51), even though they disagree whether this involves quiet (F_I = +1.60 and F_{II} = −1.33).

MAJOR CONCEPTS

In many research situations, factoring variables via R-technique addresses important issues, such as better understanding the nature of intelligence or self-concept. But both people and time can also be factored. Factoring people is often of interest in psychology, because psychological theory often is about types of people (e.g., Type A personalities).

Q-technique factor analysis is the second-most commonly used EFA method, after R-technique. In Q-technique, deciding how many people factors to extract addresses the question "How many types of people are there?" across the variables used in a given analysis. Persons' pattern or structure coefficients large in magnitude (plus or minus) on a given factor can be examined to determine "Which factored people belong to a given type?" And the factor scores that are large in magnitude can be examined to determine "Which variables were the basis for delineating the different person factors?"

8

CONFIRMATORY ROTATION AND FACTOR INTERPRETATION ISSUES

There are various situations in which the researcher may want to evaluate how well two different factor structures fit each other. Just as unrotated factors cannot be readily interpreted, as reflected in the Table 3.6 results (see chap. 3), factors across samples often cannot be accurately compared unless the two structures are compared *analytically* using Procrustean rotation methods.

These methods rotate one structure to a "best-fit" position with a second target structure. If the two factor matrices do not match under these conditions, they cannot fit under any other conditions, and the structures simply are incompatible. Some example applications of these methods include empirically evaluating

1. whether the factor pattern/structure matrices for a given set of variables across different samples are invariant across samples;
2. whether there are substantively based differences in structure for a set of items across groups (e.g., graduate students vs. faculty);
3. whether the structure is replicable using "internal replicability" (Thompson, 1994, 1996) cross-validation methods (i.e., randomly splitting the sample to inform judgments about replicability); and

4. the "factor adequacy" of an obtained pattern/structure matrix by evaluating its fit to a theoretically expected target structure consisting of ones (or negative ones) and zeroes (Thompson & Pitts, 1982).

Thompson (1992b) provided a partial statistical significance test distribution for evaluating these factor correlations across samples.

"BEST-FIT" ROTATION EXAMPLE

For heuristic purposes, the invariance of the principal components varimax-rotated pattern/structure coefficients involving perceptions of library service quality with regard to 11 variables (PER1 through PER11) will be evaluated. The comparison involves the structure for 100 faculty reported in Table 5.5 (see chap. 5) being compared with the corresponding structure of 100 graduate students (see Appendix A). The graduate students' structure will be used as the target matrix for this analysis.

Veldman (1967, pp. 238–244) explained the mathematical theory underlying the method. He noted that the cosines among the factors across the two solutions equal the factor correlations across samples. He also noted that this matrix of factor cosines can be used to rotate a second matrix to its best-fit position with a target matrix.

The cosine matrix is computed using the matrix algebra formula:

$$C_{J \times I} = (A'_{I \times K} \, B_{K \times J})'_{J \times I} \, V_{I \times J} \, E_{\Delta \times J}^{-1.5} \, V'_{J \times I},$$

where I = the number of factors in the A target pattern matrix; J = the number of factors in the B pattern matrix; and K in the number of variables in the two analyses. This algorithm requires first the computation of

$$Q_{I \times I} = A'_{I \times K} \, B_{K \times J} \, (A'_{I \times K} \, B_{K \times J})'_{J \times I}.$$

The matrix $V_{I \times J}$ is the matrix of pattern coefficients extracted from $Q_{I \times I}$. The matrix $E_{\Delta \times J}$ has zeroes off the diagonal and on the diagonal the eigenvalues of $Q_{I \times I}$.

This analysis can be conducted in SPSS using the syntax presented in Appendix B. By the way, this syntax invokes the MATRIX procedures available in SPSS, if one learns how to type syntax rather than executing SPSS via point and click. In Appendix B, all the syntax between the MATRIX command and the END MATRIX command executes matrix algebra. The adventurous reader can access an SPSS manual to discover how to command SPSS to execute any matrix procedure (e.g., transpose, inversion, multiplication) or matrix formula.

TABLE 8.1
Factor Correlations (Cosines)

Student target	Faculty matrix		
	FACT_IB	FACT_IIB	FACT_III
FACT_IA	−.007	1.000	−.024
FACT_IIA	1.000	.006	−.022
FACT_III	.022	.024	.999

Table 8.1 presents the factor correlations across the two samples. The results indicate that the factors were highly congruent or invariant across samples (e.g., $r_{IA \times IIB} = 1.00$), although the order of the factors was different in the two samples (e.g., Factor I in the student sample was Factor II in the faculty sample).

Table 8.2 presents the factor pattern/structure matrix of the 100 faculty after rotation to "best fit" with the structure of the 100 graduate students. The structures are very similar across the two groups in this data set.

The program also produces cosines (i.e., here correlations) between the variables in the target pattern/structure matrix and the variables in the best-fit rotated pattern/structure coefficient matrix. These cosines among variables should all be reasonably large to indicate that the two solutions fit reasonably well at the level of each of the variables, and not just at the factor level. If some variables' cosines are smaller, these are the variables that could not be related well in the best-fit rotation.

For the Table 8.2 data, the lowest of the 11 variables' cosines was .94 for variable PER9, and the 10 remaining cosines were all .98 or larger. These results confirm factor fit even across the 11 measured variables.

TABLE 8.2
Pattern/Structure Matrix for 100 Faculty Rotated to "Best-Fit" Position With the Pattern/Structure Matrix for 100 Graduate Students

Variable	Factor		
	FACT_I	FACT_II	FACT_III
PER1	.890	.204	.213
PER2	.868	.109	.239
PER3	.726	.231	.185
PER4	.866	.164	.266
PER5	.115	.904	.240
PER6	.124	.892	.239
PER7	.167	.905	.223
PER8	.315	.792	.170
PER9	.087	.323	.677
PER10	.209	.166	.801
PER11	.321	.105	.730

FACTOR INTERPRETATION CONSIDERATIONS

These results afford an opportunity to state some general principles of result interpretation. The statement of these principles requires specifying which aspects of results are and are not relevant to the interpretation process.

Factor Order

Both before and after rotation, factors are ordered by the amount of information they contain. Prior to rotation, the eigenvalues characterize the amount of information contained in the *unrotated* factors. For example, if the first eigenvalue associated with the first principal component was 3.0, and 11 variables were included in the R-technique analysis, the result indicates that 27.2% (3.0 / 11) of the variance in the variables can be reproduced by the first component.

Factor rotation redistributes the variance across the factor in service of pursuing simple structure. The eigenvalues are *not* relevant to evaluating the information contained in particular rotated factors. Computer packages print sums of squares to address the issue of the information contained within specific rotated factors.

In general, the order of the factors is not of interest in evaluating the solution. For example, here the analyst expected three factors in both the student and the faculty samples: "Service Affect," "Information Access," and "Library as Place." These dimensions did emerge in both samples. No particular factor ordering was theoretically expected in this analysis. It does not matter that the factors emerged in a different sequence in these analyses.

In some analyses there might be an expectation that one factor will dominate the factor space (i.e., contain a hugely disproportionate amount of variance). Only in such cases would a theoretical expectation regarding factor ordering be relevant.

"Reflection" of Factors

Some factors involve pattern and structure coefficients with different signs. Because most people find it easier to think in positive terms, if the larger pattern and structure coefficients have negative signs on a given factor, it is completely appropriate for the analyst to "reflect" the factor by reversing the signs of the coefficients on any given factor (Gorsuch, 1983, p. 181).

This is appropriate because psychological variables typically have no intrinsic scaling direction. For example, an achievement test could be scored by counting either the number of right answers or the number of wrong answers.

However, when a given factor is reflected, *all* the signs of all the pattern and the structure coefficients on the factor must be reversed. It is not necessary to acknowledge the reflection process in describing results. And in a given solution, any combination of different factors may be reflected, including, of course, none.

Salience of Variables

It is common to see researchers invoking some criterion to determine what variables are salient to the interpretation of a given factor. Values of pattern/structure coefficients of |.3|, |.35|, and |.4| are used with some frequency. Such values may be useful in setting the midpoint of gaps between smaller and larger coefficients on a given factor.

Researchers may underline or italicize the entries in pattern/structure matrices that meet their criteria, to help readers see patterns. Or researchers may sort the rows of the matrices by the values of the biggest absolute coefficients on different factors.

Some researchers "blank out" nonsalient coefficients to make simple structure more apparent. Unlike italicizing coefficients or sorting factor rows, omitting nonsalient coefficients is *not* good practice. Doing so deprives other researchers of the opportunity to conduct secondary analyses, such as invoking different rotation methods, or conducting best-fit rotation of their results with yours.

It is important not to get too rigid in determining salience criteria. Use a salience criterion in a given study that does a good job for the given results in highlighting simple structure.

Naming Factors

An important part of the interpretation process is naming the factors. This is where the art of the factor analytic process is exercised. Factor names are usually one- or two-word phrases.

Factors should *never* be named after measured variables. A factor is inherently an aggregation of several measured factors. A factor reflected in only a single measured variable (i.e., a "specific" factor) is inherently not a true factor. Thus, all real factors involve multiple variables, and to this extent must be named in a manner reflecting the overall pattern of contribution of different variables to the factor's definition (or conversely reflecting the underlying construct's manifestation via the measured variables).

Factor naming also must be informed by the differential saturation of a given factor by given factored entities. For example, in an orthogonal R-technique analysis, if one variable has a pattern/structure coefficient of +.7 on a factor, and another variable has a pattern/structure coefficient of +.5, do not forget that the first variable has 49% of its variance in common with the factor, whereas the second measured variable has only 25% common variance. Thus, the first variable should have roughly double influence on selection of the factor label.

Factor names must be sufficiently broad to capture all of the primary relationships suggested by the pattern and structure coefficients. But there is another caveat that must be mentioned. When conducting factor analyses and comparing results with those in prior related studies, it is important to not place undue reliance on construct names assigned by previous researchers. Researchers may have named their factors in ways that do not seem insightful or even reasonable. If the factor labels from prior studies are accepted without scrutiny, invalid comparisons with related prior findings may result.

MAJOR CONCEPTS

In various situations researchers may want to determine how well two different factor structures for a given set of factored entities (e.g., variables if R-technique is used) fit each other. Empirical methods usually must be used for this purpose, because apparent differences may be method artifacts, and empirical methods provide a quantification of degree of structure correspondence. The Appendix B SPSS syntax can be used as a model to conduct such best-fit rotations.

In evaluating correspondence of factors across samples or subsamples, factor order is usually not a consideration. Similarly, a given factor across samples may be oriented in opposite directions, such that the coefficients for a factor in one sample must be "reflected" by changing *all* the signs of *all* coefficients on the factor. Factor orientations are completely arbitrary, but we usually prefer the predominance of large magnitude coefficients to be positive, because most people find it easier to think positively.

When presenting factor structures, researchers often highlight coefficients that they deem salient. Such salience criteria are usually set by looking for gaps between near-zero and nonzero coefficients (e.g., absolute .4) in a given solution. Salience criteria should not be used with excessive rigidity.

Factors are synthetic variables reflecting underlying latent constructs, and therefore should never be named after measured variables. Equally important, when reading prior research, do not accept the factor labels of others at face value. When conducting either subjective or empirical factor meta-analyses (Thompson, 1989), focus on the invariance of the factor meanings across studies, notwithstanding the labels assigned by different researchers.

9

INTERNAL REPLICABILITY ANALYSES

In quantitative research, all statistical methods, including factor analysis, are susceptible to the influences of three kinds of errors: sampling error, measurement error, and model specification error. As noted in chapter 4, *sampling error* is that portion of variance in sample data that is unique to a given sample, and in effect reflects the "personality" of a given sample. By definition, the sampling error in a given sample does not exist in the population.

Measurement error is the variance in data that is due to inherent imperfections in measurement (Thompson, 2003). We estimate these influences by computing ratios of reliable variance to the total observed score variance; these ratios are called *reliability coefficients*. In every study it is important to estimate the reliability of the data in hand in the study and not merely cite reliability coefficients from the manual or prior studies (Vacha-Haase, Kogan, & Thompson, 2000). As Wilkinson and the Task Force on Statistical Inference (1999) emphasized:

> It is important to remember that a test is not reliable or unreliable. . . . Thus, authors should provide reliability coefficients of the scores for the data being analyzed even when the focus of their research is not psychometric. (p. 596)

Model specification error is the failure to use all and only the correct variables, or the failure to use the correct statistical analysis, or both, thus obtaining incorrect results. Of course, as Pedhazur (1982) has noted, "The rub, however, is that the true model is seldom, if ever, known" (p. 229). And as Duncan (1975) has noted, "Indeed it would require no elaborate sophistry to show that we will never have the 'right' model in any absolute sense" (p. 101).

Sampling error is particularly pernicious in multivariate analyses, because more complex analyses afford more opportunities for these influences to be manifested. With respect to exploratory factor analysis (EFA), Gorsuch (1983) noted, "In factor analysis, one has numerous possibilities for capitalizing on chance. Most extraction procedures, including principal factor solutions, reach their criterion by such capitalization. The same is true of rotational procedures, including those that rotate for simple structure" (p. 330).

There are two major ways of empirically addressing result replicability questions: *external* replicability analyses and *internal* replicability analyses (Huberty, 1994; Thompson, 1996). External replicability studies involve collection of independent data from different participants. These are true replication studies and provide the best evidence of result replicability by directly and explicitly addressing replicability concerns.

However, many researchers find themselves unable to always conduct external replicability studies. In these situations researchers may use some combination of three major logics for internal replicability analyses: cross-validation, the jackknife, and the bootstrap (Thompson, 1993b, 1994). These three methods "combine the subjects in hand in different ways to determine whether results are stable across sample variations, that is, across the idiosyncracies of individuals that make generalization in social science so challenging" (Thompson, 1996, p. 29).

Here both EFA cross-validation and bootstrap methods will be illustrated. These are, respectively, the simplest and the most complex internal replicability methods. Jackknife applications have been discussed elsewhere (Huberty, 1994).

Across the three major internal replicability analysis choices it is important to remember that *none* of the methods is as useful as true replication studies. As Thompson (1993b) noted, because

> all three strategies are typically based on a single sample of subjects in which the subjects usually have much in common (e.g., point in time of measurement, geographic origin) relative to what they would have in common with a separate sample, the three methods all yield somewhat inflated estimates of replicability. Because inflated estimates of replicability provide a better estimate of replicability than no estimate at all (i.e., statistical significance testing), these procedures can still be useful in focusing on the *sine qua non* of science. (pp. 368–369)

Cross-validation and bootstrap applications will be illustrated here using for heuristic purposes the first 30 cases of data and variables PER1 through PER8 from Appendix A.

EFA CROSS-VALIDATION

Cross-validation methods executed as internal replicability analyses randomly split the data into two subgroups. Analyses are conducted separately within each subgroup, and then an *empirical* analysis must be executed to quantify the degree of fit across the two analyses. These methods have been used across the general linear model, from regression to canonical correlation analyses (cf. Crossman, 1996; Thompson, 1994). The methods are logically straightforward and do not require specialized software.

Table 9.1 presents the cross-validation subgroup assignments ("1" or "2") for the 30 cases produced by flipping a coin. For logistical purposes as is typically done, because keeping track of the subgroups is thereby facilitated, here the 30 cases were split 16/14 rather than 15/15. Table 9.2 presents the pattern/structure matrices from principal components analyses rotated to the varimax criterion for both subgroups.

The best-fit rotation procedure must then be invoked to align the factor structures to optimal fit. Either solution can be used as the target for this analysis, although some preference might be afforded to using the larger sample as the target. Table 9.2 also presents the second solution rotated to best-fit position with the solution for the first subgroup. In this small heuristic example the solutions in the two subgroups are reasonably congruent. This is reflected in the cosines of the two solutions virtually being an identity matrix.

The biggest disparity across the subgroups involves the variable PER8, which in the second subgroup had pattern/structure coefficients on Factor I of .667 and on Factor II of .596. In the first subgroup, PER8 had a trivial pattern/structure coefficient on Factor I ($.158^2$ = 2.5%) but a huge value on Factor II ($.840^2$ = 70.6%). This disparity is also reflected in the cosines for the eight variable pairs each being greater than .94, except for PER8, which had a cosine of .808.

Cross-validation can give some insight regarding potential result replicability. If solutions will not replicate across subgroups in an internal replicability analysis involving a single sample, then replication with a truly independent second sample seems particularly unlikely.

And it should be emphasized that

> the full sample results are always the basis for our interpretations. The subsample results are only employed as a vehicle for exploring the replicability of the full sample results. The full sample results are used

TABLE 9.1
Cross-Validation Random Subgroup Assignments and Bootstrap
Resampling Assignments

Cross-validation			Resample			
Case	Subgroup	Draw	1	2	3	4
1	2	1	1	30	20	25
2	1	2	29	4	27	19
3	2	3	18	12	26	2
4	2	4	13	7	20	17
5	2	5	24	3	17	25
6	2	6	13	25	24	24
7	2	7	17	17	23	4
8	2	8	15	17	1	23
9	1	9	11	28	18	8
10	1	10	16	6	29	12
11	2	11	10	24	6	13
12	1	12	22	10	3	1
13	1	13	21	19	10	8
14	1	14	26	30	13	27
15	1	15	13	1	15	11
16	1	16	24	20	15	20
17	1	17	5	5	28	1
18	1	18	12	17	4	17
19	1	19	22	21	19	30
20	2	20	15	24	3	8
21	2	21	1	10	24	22
22	1	22	26	6	26	15
23	1	23	14	28	11	1
24	2	24	15	15	9	1
25	2	25	24	10	19	19
26	1	26	28	12	1	3
27	2	27	20	5	4	7
28	2	28	24	28	16	28
29	1	29	22	25	19	5
30	1	30	25	23	18	11

> for interpretation, because we tend to vest more confidence in results
> as sample size increases, and the full sample analysis involves more
> subjects than do the subgroup analyses. (Thompson, 1994, p. 172)

The subgroups are only used as a vehicle to evaluate the possible replicability
of the factor solution for the complete sample.

EFA BOOTSTRAP

The "bootstrap" is an elegant computer-intensive internal replicability
analysis conceptualized for several analyses by Efron and his colleagues.
Diaconis and Efron (1983) provide an accessible article summarizing this

TABLE 9.2
Varimax-Rotated Factor Pattern/Structure Matrix for Two Subgroups
($n_1 = 16$; $n_2 = 14$)

Variable	Subgroup #1 (target) I	II	Subgroup #2 I	II	Subgroup #2 after best-fit I	II
PER1	.896	.340	.929	.214	.923	.238
PER2	.954	.110	.942	.250	.935	.274
PER3	.784	.474	.809	.496	.796	.516
PER4	.877	.354	.596	.487	.583	.502
PER5	.376	.792	.345	.898	.322	.906
PER6	.326	.845	.356	.890	.333	.899
PER7	.261	.906	.228	.926	.205	.932
PER8	.158	.840	.667	.596	.651	.613

Note. Pattern/structure coefficients greater than |.5| are presented in italics.

logic in layperson's terms. Lunneborg's (2000) book provides a more technical and comprehensive view.

The bootstrap can be conceptualized as a method of using the sample data in such a manner that the sample data are themselves used to estimate the distribution of a large number of sample results from the population (i.e., the sampling distribution). This process of the estimation pulling itself up by its own bootstraps explains why the approach is named the bootstrap.

One way of explaining the technique, although somewhat simplified, begins by describing the creation of a "mega-file" created by copying the sample data set of size *n* onto itself (i.e., concatenating) a very large number of times. Then repeated resamples of cases, each of exactly size *n*, are drawn from the mega-file, and in each resample the results are analyzed independently.

The number of resamples is usually at least 1,000, although 2,000 to 5,000 resamples are recommended. In a given resample, a given case might be drawn not at all, once, or several times. And this case might be drawn not at all, once, or several times in the next resample. After all the resamples are drawn, mean (or median) parameter estimates (e.g., R^2, factor pattern/structure coefficients) are computed across the resamples.

The standard deviations of the estimates are also computed. These are *empirical* (as against theoretically based) estimates of the standard errors of each of the parameter estimates.

There is only one difficulty in applying the bootstrap in multivariate statistics such as EFA: When more than one set of weights (e.g., factors) can be computed, the weights across the resamples must be rotated to best fit with a *common* target matrix prior to computing averages and standard errors. This is because in the first resample a construct might emerge as Factor I, but in the next resample might emerge as Factor II. Factors must

TABLE 9.3
First Four Eigenvalues From Four Resamples ($n = 30$)

Resample	Factor		
	I	II	III
1	3.867	2.523	0.600
2	4.798	1.776	0.675
3	5.357	1.312	0.548
4	5.190	1.553	0.605
Mean	4.803	1.791	0.607
SE/SD	0.577	0.454	0.045

be rotated to best-fit position so that averages of apples and apples are being computed across the resamples.

Of course, these mechanics are not difficult when they are automated, and bootstrap applications across various multivariate analyses have been reported and recommended (Thompson, 1988, 1992a, 1995). In EFA the bootstrap can be used to inform the decision of how many factors to extract, to help evaluate the replicability of the structure, or both.

Factor Extraction Bootstrap Procedures

The bootstrap can be used as another procedure to decide how many factors to extract, in addition to the methods described in chapter 3. In this heuristic example only four resamples are drawn. This makes the summary manageable, but of course is not a realistic description of an actual bootstrap analysis.

As reported in Table 9.1, in the first resample, the first case drawn was Case #1. In each draw *all* the data for a given case are extracted into the resample. Case #1 was also drawn again in Draw 21.

In each resample the eigenvalues are computed. Then means of the eigenvalues and the standard deviations can be derived. Because the resamples are being used to estimate sampling distributions, these standard deviations are actually empirically estimated standard errors (i.e., SDs of the parameter estimates in the sampling distribution).

Table 9.3 presents the first three of the eight eigenvalues from principal components analyses of the data from the four resamples. The table also presents the mean eigenvalues across resamples and the standard errors of these estimates. For example, the mean eigenvalue across four resamples for the second factor without rotation was 1.791 ($SE_\lambda = 0.454$). The mean eigenvalue across four resamples for the third factor was 0.607 ($SE_\lambda = 0.045$).

For these results the extraction of two factors is suggested. If the third eigenvalue mean had been 0.999, and the standard error had been small but nontrivial, the extraction of the third factor might have been warranted. In other cases, even if an eigenvalue mean is slightly greater than 1.0, a relatively large standard error would suggest that not much faith can be vested in this eigenvalue consistently being greater than 1.0; here some thought might be given to not extracting the factor associated with this relatively small and unstable eigenvalue.

Pattern/Structure Coefficients and Standard Errors

Bootstrap estimation of factor pattern/structure coefficients and their standard errors requires the specification of a common factor space or target to which all resampled solutions can be rotated to a position of best fit. The structure in the actual sample can be used as the target for this purpose.

Another choice is a theoretically expected factor structure. In the present example, the target matrix reflected an expectation that variables PER1 through PER4 would have +1.0 coefficients on Factor I, and zeroes on Factor II. The target matrix also reflected an expectation that variables PER5 through PER8 would have +1.0 coefficients on Factor II, and zeroes on Factor I.

Table 9.4 presents the varimax-rotated pattern/structure matrices from principal components analyses in the four resamples. Note that the first factor includes variables PER1 through PER4 in Resamples #2 and #3 but not #1 and #4.

Table 9.5 presents the four structures each rotated to best-fit position with the target matrix. Table 9.6 summarizes the means of the parameter estimates and their standard errors. Table 9.6 also reports the mean parameter estimates divided by their respective standard errors.

The ratio of a parameter estimate to its standard error (SE) is very important in statistics, and so this ratio is given many names. These names include *critical ratio*, *Wald statistic*, and *t*. We usually expect these *t* values to be greater than roughly |2.0| for parameters that we wish to interpret as being nonzero.

Overall, the heuristic results, if they had been real, would have supported a conclusion that the factor structure was reasonably replicable. This is reflected in the fact that the empirically estimated standard errors for large values in Table 9.6 are quite small (e.g., .039, .009). These standard errors indicate that the noteworthy parameters in the model do not vary much across various configurations of the people considered in the bootstrap resamples. If we mix up our sample cases in many different configurations and always still get the same basic factor structure, this would give some hope that the structure may also replicate across the configuration of participants in a new sample.

TABLE 9.4
Varimax-Rotated Pattern/Structure Matrix From Four Resamples ($n_K = 30$)

Variable	Factor I	Factor II
Resample #1		
PER1	−.066	.921
PER2	−.067	.958
PER3	.347	.836
PER4	.251	.808
PER5	.847	.219
PER6	.904	.031
PER7	.917	−.021
PER8	.797	.135
Resample #2		
PER1	.966	.136
PER2	.964	.130
PER3	.927	.266
PER4	.759	.431
PER5	.128	.927
PER6	.344	.784
PER7	.182	.845
PER8	.159	.775
Resample #3		
PER1	.869	.297
PER2	.947	.127
PER3	.810	.493
PER4	.859	.292
PER5	.555	.670
PER6	.226	.899
PER7	.251	.882
PER8	.228	.825
Resample #4		
PER1	.067	.947
PER2	.215	.933
PER3	.485	.836
PER4	.541	.632
PER5	.922	.203
PER6	.911	.228
PER7	.878	.148
PER8	.788	.329

MAJOR CONCEPTS

All statistical methods, including factor analysis, are susceptible to the influences of three kinds of errors: sampling error, measurement error, and model specification error. To address the influences of sampling error, researchers replicate results with independent samples (i.e., external replica-

TABLE 9.5
Four Resample Solutions Rotated to Best Fit With Target

	Factor	
Variable	I	II
Resample #1		
PER1	.920	−.079
PER2	.957	−.080
PER3	.841	.335
PER4	.812	.240
PER5	.231	.844
PER6	.044	.904
PER7	−.008	.917
PER8	.146	.795
Resample #2		
PER1	.968	.125
PER2	.966	.119
PER3	.930	.255
PER4	.764	.422
PER5	.139	.926
PER6	.354	.779
PER7	.192	.843
PER8	.168	.773
Resample #3		
PER1	.864	.312
PER2	.944	.143
PER3	.802	.507
PER4	.854	.306
PER5	.544	.679
PER6	.211	.903
PER7	.236	.886
PER8	.214	.828
Resample #4		
PER1	.950	.012
PER2	.944	.159
PER3	.863	.435
PER4	.663	.503
PER5	.256	.909
PER6	.282	.896
PER7	.199	.868
PER8	.375	.767

tion). When external replication is impossible or impractical, internal replicability methods, such as cross-validation or the bootstrap, may be useful.

Exploratory factor analysis cross-validation methods can be conducted
by randomly splitting the sample and using SPSS to factor analyze both
data sets, and then using the Appendix B best-fit rotation to quantify degree
of factor fit across the two subsamples. However, the basis for actual factor
interpretation is always the structure from the total sample. The subsamples

TABLE 9.6

Bootstrapped Estimates of Pattern/Structure Coefficients and Their Standard Errors

Variable	Factor I			Factor II		
	Mean	SD/SE	t	Mean	SD/SE	t
PER1	.926	.039	23.47	.093	.146	0.63
PER2	.953	.009	102.33	.085	.096	0.88
PER3	.859	.046	18.49	.383	.096	3.99
PER4	.773	.071	10.86	.368	.102	3.62
PER5	.293	.152	1.93	.840	.098	8.60
PER6	.223	.115	1.94	.871	.053	16.45
PER7	.155	.095	1.62	.879	.027	32.58
PER8	.226	.090	2.52	.791	.024	33.09

Note. Pattern/structure coefficients greater than |.4| are italicized.

are created only to gain insight regarding the character of the total data set and *not* for factor interpretation.

The EFA bootstrap (Thompson, 1988) is a particularly powerful internal replicability method. The bootstrap is so powerful because it resamples the data in so many different combinations. However, as a practical matter, specialized software must be used to conduct these analyses.

10

CONFIRMATORY FACTOR ANALYSIS DECISION SEQUENCE

Although the basic concepts of factor analysis are roughly a century old (see Spearman, 1904), more recent extensions of the exploratory factor analysis (EFA) concepts have created the basic methods for confirmatory factor analysis (CFA; see Jöreskog, 1969). Both EFA and CFA are part of the general linear model (GLM). Thus, many of the concepts introduced previously with regard to EFA apply in CFA as well. Indeed, an important emphasis in this book is elaborating the similarities between EFA and CFA as part of a single underlying GLM, while also highlighting the differences in the two sets of methods.

Of course, the basic difference in CFA is that one or more underlying models (i.e., how many factors, which measured variables are thought to reflect which latent variables, whether the factors are correlated) *must* be specified even to run the analysis. There are also statistical analyses that are possible in CFA but impossible in EFA (e.g., allowing error variances to be correlated).

And some procedures that are routine in EFA, such as factor rotation, are irrelevant in CFA. This is because the a priori models themselves typically specify simple structure, by constraining certain factor pattern coefficients to be zero while freeing other pattern coefficients to be estimated.

Confirmatory factor analysis is a very important component within a broader class of methods called *structural equation modeling* (SEM), or sometimes *covariance structure analysis*. CFA specifies the "measurement models" delineating how measured variables reflect certain latent variables. Once these measurement models are deemed satisfactory, then the researcher can explore path models (called *structural models*) that link the latent variables.

Structural equation modeling is beyond the scope of the present volume. Excellent book-length treatments of SEM are available elsewhere (see Byrne, 1994, 1998, 2001). Chapter-length expositions are also available (see Thompson, 2000c). Schumacker and Marcoulides (1998) and Marcoulides and Schumacker (2001) describe recent extensions of these methods.

Suffice it to say that even when the researcher is applying the wider range of analyses possible with SEM, CFA models are often a critical starting place for such analyses (Thompson, 2000c). It makes little sense to relate constructs within an SEM model if the factors specified as part of the model are not worthy of further attention.

CONFIRMATORY FACTOR ANALYSIS VERSUS EXPLORATORY FACTOR ANALYSIS

In EFA, all parameters implicit in a factor model *must* be estimated. For example, if there are 9 measured variables, and 3 principal components are extracted and then rotated to the promax criterion, then 27 (9×3) pattern coefficients, 27 (9×3) structure coefficients, 9 communality coefficients (h^2; or their corresponding uniquenesses estimated as $u^2 = 1 - h^2$), and 3 factor correlation coefficients ($r_{I \times II}$, $r_{I \times III}$, $r_{II \times III}$) are estimated.

Of course, because $\mathbf{P}_{9 \times 3}$ $\mathbf{R}_{3 \times 3} = \mathbf{S}_{9 \times 3}$, given $\mathbf{P}$, if either the factor correlation matrix ($\mathbf{R}_{3 \times 3}$) or the factor structure matrix ($\mathbf{S}_{9 \times 3}$) is mathematically determined, so is either $\mathbf{R}$ or $\mathbf{S}$, and thus in actuality only 30 parameters $(27 + 3)$ are estimated. Note that the uniquenesses are also mathematically determined once the pattern and structure matrices are estimated.

In CFA, the researcher can "constrain" or "fix" certain parameters to mathematically "permissible" values (e.g., a variance may be constrained to equal any positive number; a correlation, r, may be constrained to equal -1, $+1$, or any number in between), while at the same time "freeing" the use of the input data to derive estimates of other model parameters (e.g., factor pattern coefficients, factor variances). For example, the researcher might constrain variables V1, V2, and V3 to reflect only Factor I, variables V4, V5, and V6 to reflect only Factor II, and variables V7, V8, and V9 to reflect only Factor III, with the three factors allowed to be correlated.

In EFA, the researcher may expect certain coefficients, but these expectations *cannot* be incorporated into the analysis. In CFA a researcher *must* declare as *input* into the analysis one or more specific models, each containing some "fixed" and some "freed" parameters.

The original computer software for CFA (and SEM), such as LISREL (i.e., analysis of LInear Structural RELationships; Jöreskog & Sörbom, 1984), required the researcher to input these model expectations using matrix algebra. A thorough knowledge of both the uppercase and the lowercase Greek alphabet was also required!

Modern software, such as AMOS (Arbuckle & Wothke, 1999) and EQS (Bentler, 1995), instead uses a graphics-oriented user interface. The researcher's expectations are declared using the software to draw a diagram representing the input model. The results are then reported by the software in text form (i.e., sets of coefficients printed for different matrices), but also as an output diagram with estimates printed on the input model diagram.

By tradition, in these various computer programs measured variables are represented by squares or rectangles. Latent variables are represented as circles or ovals. Regression paths or, in the factor analysis case, pattern coefficients, are represented by one-headed arrows. These are drawn *from* factors *to* measured variables to reflect the premise that the latent variables are in fact an underlying influence on the manifestation of the factors in the form of scores on the measured variables. Correlations (and covariances) are represented as two-headed arrows drawn to connect either measured or latent variables in pairs for which correlations or covariances are freed to be nonzero and are estimated in the analysis.

Figure 10.1 presents the input graphic model declaration implied by an EFA involving nine measured variables and three correlated factors. Table 10.1 presents a sequential count of the 30 parameters that are estimated as well as a sequential count of the parameters mathematically determined once the estimates are in hand.

Figure 10.2 presents the related CFA input model presuming that each of the nine measured variables is associated with only one factor, three measured variables per factor. The absence of paths between selected pairs of measured and latent variables (e.g., between V1 and Factor II, between V2 and Factor II) implies that these factor pattern coefficients have been constrained to be zero. This model declares simple structure on its face, because no measured variable is allowed to function as an indicator for more than one factor. Thus, if model fit is adequate, no rotation will be necessary.

Table 10.2 presents a sequential count of parameters estimated, not estimated, and implied by this model. However, one difference between EFA and CFA arises in this presentation. Note from Table 10.2 that the nine uniquenesses (u^2) are counted as estimated in the CFA analysis. In the EFA analyses these were taken as implied, because once the pattern and structure coefficients are in hand, the communality coefficients (h^2) are

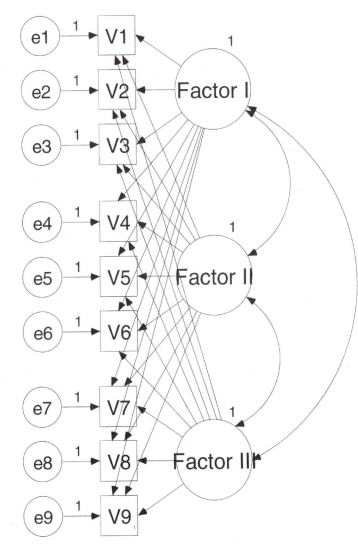

Figure 10.1. Exploratory factor analysis factor model with nine measured variables and three correlated factors.

computed by simple multiplications and additions along a given row of the factor matrix, and then uniquenesses are computed as $u^2 = 1 - h^2$.

Figure 10.3 presents yet another factor model for this research problem. This figure illustrates some of the model elements that are not possible in EFA. First, the uniquenesses or error variances involving "e1" and "e2" have been allowed to covary or be correlated in this model. No error variances can be correlated in an EFA model, but in CFA the correlations of various

TABLE 10.1
Counts of Parameters Estimated (and Implied) by the Figure 10.1
Factor Model

Variable	Pattern			Structure			u^2
	I	II	III	I	II	III	
V1	1	10	19	(31)	(40)	(49)	(58)
V2	2	11	20	(32)	(41)	(50)	(59)
V3	3	12	21	(33)	(42)	(51)	(60)
V4	4	13	22	(34)	(43)	(52)	(61)
V5	5	14	23	(35)	(44)	(53)	(62)
V6	6	15	24	(36)	(45)	(54)	(63)
V7	7	16	25	(37)	(46)	(55)	(64)
V8	8	17	26	(38)	(47)	(56)	(65)
V9	9	18	27	(39)	(48)	(57)	(66)
Factor r							
I	—						
II	28	—					
III	29	30	—				

Note. Coefficients mathematically determined and thus implied once estimates are in hand are noted in parentheses.

pairs of error variances can be estimated, as deemed necessary by the researcher.

Second, only two factors are allowed to be correlated, and this correlation is estimated as part of the model. In EFA, factors must be either all correlated or all uncorrelated.

Third, an "equality" constraint, designated by the same letter (here *a*) being placed on the paths for the factor pattern coefficients for three measured variables on Factor III, has been imposed. This means that the input data will be consulted to estimate these three pattern coefficients, but the estimate has been constrained such that the same single number *must* be used for all three of these pattern coefficients. Equality constraints are not possible within EFA.

Equality constraints have several possible uses in EFA, but one use is to test a theoretical expectation that all measured variables on a given factor measure that factor equally well. For example, when scoring on test subscales is done by summing item scores on a given subscale, the implication is that the items on the subscale measure the underlying construct equally well. This premise can be directly tested in CFA.

In CFA one may also (and should!) test rival models. Equality constraints might be imposed in one model; however, in a rival model all freed pattern coefficients might be estimated with no equality constraints. This rival model might reflect a belief that the test items do reflect the intended

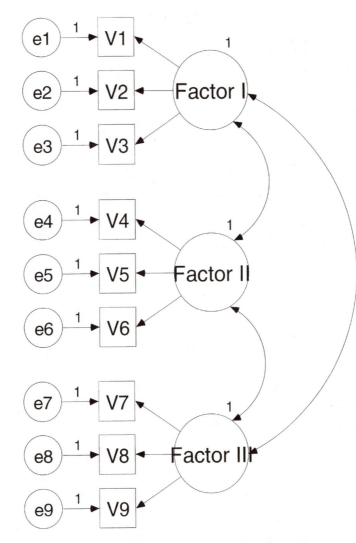

Figure 10.2. Confirmatory factor analysis factor model with nine measured variables and three correlated factors.

constructs, but perhaps not equally. The differential fit of these rival models with and without equality constraints can then be quantified.

PREANALYSIS DECISIONS

In CFA, as in EFA, a sequence of decisions must be made when conducting an analysis. Some of the decisions are the same as those made in EFA, but some are unique to the confirmatory rubric.

TABLE 10.2

Counts of Parameters Estimated, Not Estimated, (and Implied) by the
Figure 10.2 Factor Model

Variable	Pattern			Structure			u^2
	I	II	III	I	II	III	
V1	1	—	—	(22)	(31)	(40)	13
V2	2	—	—	(23)	(32)	(41)	14
V3	3	—	—	(24)	(33)	(42)	15
V4	—	4	—	(25)	(34)	(43)	16
V5	—	5	—	(26)	(35)	(44)	17
V6	—	6	—	(27)	(36)	(45)	18
V7	—	—	7	(28)	(37)	(46)	19
V8	—	—	8	(29)	(38)	(47)	20
V9	—	—	9	(30)	(39)	(48)	21
Factor r							
I	—						
II	10	—					
III	11	12	—				

Note. Coefficients mathematically determined and thus implied once estimates are in hand are noted in parentheses. "Fixed" parameters are designated by dashes ("—"). Here 18 factor pattern coefficients are constrained to be zeroes. And the three factor variances are fixed to equal 1.0s, thus making the factor covariances equal factor correlation coefficients, because for example $r_{I \times II} = COV_{I \times II} / (1.0 \times 1.0)$.

Rival Models

Like SEM, CFA involves testing the fit of models to data. When models fit well, as reflected in various statistical fit indices, some researchers erroneously infer that the correct model has been specified, and that in some sense the model is proven.

In fact, it is conceivable that several models may fit a given data set. As Thompson (2000c) noted, "the fit of a single tested model may always be an artifact of having tested too few models" (p. 278). Thus it is desirable to test the fit of several rival models when conducting a CFA.

The fit of a preferred model is more impressive when that fit occurs in the context of testing several rival models, especially when some of the rival models are theoretically plausible and are not articulated so as to create straw person arguments. The fit of a model is also more impressive when the tested model is strongly *disconfirmable*.

Disconfirmability is partially a function of the degrees of freedom for the model test. Degrees of freedom indicate, given the presence of one or more sample statistics (e.g., the mean of the sample data, the r between two measured variables), how many pieces of information in the analyses remain free to vary after the estimate is in hand. For example, in the presence of knowledge that the sample mean is 3, given *any* two scores in the data set (1, 3, 5), the researcher can always accurately determine the third score.

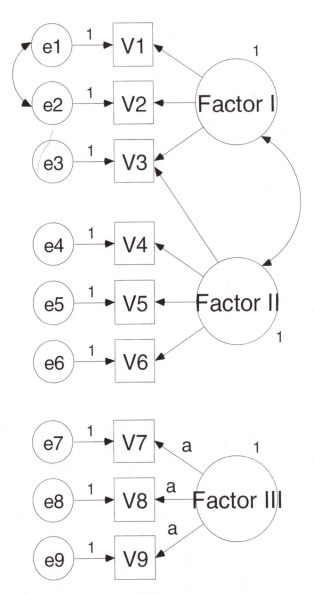

Figure 10.3. Confirmatory factor analysis (CFA) factor model with three model features unique to CFA.

For this problem the degrees of freedom is a function of the sample size, n, and equals $n - 1$.

Table 10.3 presents the computations for statistical significance testing in any r^2 or multiple regression (R^2) analysis involving p predictor variables (e.g., for the bivariate correlation, $p = 1$). There are two related situations in which $F_{\text{CALCULATED}}$ (and the corresponding $p_{\text{CALCULATED}}$ also) is unde-

TABLE 10.3
Statistical Significance Calculations for r^2 and R^2 Problems

Source	Sum-of-squares	Degrees of freedom	Mean square	$F_{\text{CALCULATED}}$	$r^2/$ R^2
Model	A	D	G	J	K
Unexplained	B	E	H		
Total	C	F	I		

Note. D = p (i.e., the number of predictor variables). F = $n-1$. E = $n-1-p$. G = A / D. H = B / E. I = C / F. J = G / H. K = A / C.

fined and a statistical significance test cannot be conducted. First, regardless of sample size, if the sum-of-squares$_{\text{UNEXPLAINED}}$ is zero (and correspondingly the r^2 or R^2 is 100%), the F cannot be calculated, because the mean square$_{\text{UNEXPLAINED}}$ will be zero, and division by zero is mathematically impermissible.

Second, whenever the degrees of freedom$_{\text{MODEL}}$ equals the available degrees of freedom (i.e., for these analyses $n-1$), the sum-of-squares$_{\text{UNEXPLAINED}}$ will always be zero. For example, if X and Y are both variables (i.e., $SD_X > 0$ and $SD_Y > 0$), and $n = 2$, r^2 must be 100%, and the sum-of-squares$_{\text{UNEXPLAINED}}$ must by definition both be zero. Similarly, in a regression analysis involving *any* two predictors, the R^2 will *always* be 100% whenever $n = 3$.

Clearly, an R^2 of 100% is entirely unimpressive when 10 predictor variables are used and n is 11! However, an R^2 of 100% would be very impressive (even though the result still could not be tested for statistical significance!) if $n = 1,000$, because this regression model would be highly *disconfirmable* given that the residual degrees of freedom$_{\text{UNEXPLAINED}}$ equaled 989.

The residual degrees of freedom are an index of the potential disconfirmability of any model. The good fit of models is more impressive when degrees of freedom$_{\text{UNEXPLAINED}}$ are large, because as a function of the residual degrees of freedom the model is parsimonious and yet has good explanatory ability.

In CFA (and SEM), however, total degrees of freedom for tests of model fit are *not* a function of the sample size, n. Instead, the total degrees of freedom in CFA are a function of the number of the unique entries in the covariance matrix, which is a function of the number of measured variables (v) in the analysis. The total degrees of freedom for these tests equals ([v ($v + 1$)] / 2). For example, in a study involving $v = 9$ measured variables, the available degrees of freedom equal ([9 (9 + 1)] / 2) = ([9 (10)] / 2) = 45. This corresponds to evaluating 9 variances and 36 covariances (i.e., there are 9 + 36 = 45 unique entries in the associated variance/covariance matrix).

In CFA (and SEM), each parameter (e.g., factor pattern coefficient, model error variance, factor correlation) that is estimated "costs" one degree of freedom. This means that *each and every* model involving nine measured variables and estimation of *any* 45 parameters will always perfectly reproduce or fit the data.

The disconfirmable and/or plausible models for a CFA are context-specific, and some such models can only be developed by consulting theory and related empirical research. However, several rival CFA models may often be useful across a fairly wide array of problems:

1. *Independence model*. This model specifies that the measured variables are all perfectly uncorrelated with each other, and thus that no factors are present. The model estimates only the model specification variances of each of the measured variables. This model is disconfirmable but usually not plausible. However, the statistics quantifying the fit of this model are very useful in serving as a baseline for evaluating the fits of plausible rival models.

2. *One-factor model*. This model is disconfirmable, but usually not plausible for most research situations. Again, however, even if this model is not theoretically plausible, the fit statistics from this model test can be useful in characterizing the degree of superiority of a rival multifactor model. And it is desirable to rule out the fit of this model even when multifactor models are preferred, so that support for the multifactor model is then stronger.

3. *Uncorrelated factors model*. In EFA, as noted in chapter 3, orthogonal rotations are the default and usually provide simple structure. However, in CFA, correlated factors are usually expected and almost always provide a better fit to the data. However, it is helpful to quantify the degree of superiority of model fit for a correlated versus uncorrelated factors model. Estimating the factor correlations for f factors costs $(f[f-1])/2$ degrees of freedom, for example, $(3[3-1])/2 = (3[2])/2 = 3$). Uncorrelated factor models are more parsimonious when there are two or more factors. The question is whether the sacrifice of parsimony (and of related disconfirmability) is worthwhile in service of securing a better model fit.

Model Identification

A model is said to be "identified" when, for a given research problem and data set, sufficient constraints are imposed such that there is a single set of parameter estimates yielded by the analysis. If for a given model an infinite number of estimates for parameters being approximated are equally plausible, the parameters are mathematically indeterminate and the model is "underidentified."

A necessary but insufficient condition for model identification is that the number of parameters being estimated is not more than the available degrees of freedom. For example, if the model involves nine measured variables ($df = 45$), the model is not identified if the model requires estimation of 50 parameters.

Model identification also requires that the measurement scale (i.e., the variance or the standard deviation) of each latent variable is specified or constrained. This is because latent variables, by definition, have no intrinsic scaling, and so there are infinitely many plausible scales for these variables, each suggesting a corresponding plausible set of estimates for the other model parameters.

In effect, if we want to estimate scores of a latent variable, we must first declare a metric for the estimate. It is usually irrelevant what this metric is (e.g., pennies, dollars, pesos, lira, pounds). But some metric must be selected. There are two common ways to identify CFA models.

First, any factor pattern coefficient on each factor can be fixed to any number. The number "1" is a common choice. But we could use any number. In effect, this says we want to scale the scores on the latent variable as some multiple of the selected measured variable.

For example, if we had three people—Carol, Colleen, and Wendy— we could measure the height of everyone using an object such as a ruler. Or we could measure everyone's height using any one of the three people as the "ruler." If Colleen was selected to identify the model, and her value as an index of height was fixed as a "1," and Wendy was 1.2 times taller than Colleen, in effect we declare Wendy's height to be 1.2 "Colleens."

The decision of which measured variable's pattern coefficient on a factor we select to fix to some number (usually "1") makes no real difference. However, some researchers stylistically prefer for model identification purposes to pick the measured variable thought to most reflect the factor or to have scores that are the most reliable from a measurement point of view.

Second, when we have a first-order factor model we can instead constrain the factor variances to be any mathematically plausible number (i.e., positive). When this strategy is selected, it is useful to use the same number to constrain all factor variances, and usually the number "1" is used for these purposes. One advantage of doing so is that the covariances of the factors then each become factor correlation coefficients, because when computing the factor correlations as $r = COV / (SD \times SD)$ all pairs of computations will then have denominators of one. This approach also has the advantage that the pattern coefficients for a factor can then be compared apples to apples as to how well each measured variable reflects the underlying construct.

The selection of scaling for the latent variables will not affect the model fit statistics. The scaling decisions are basically arbitrary and can reasonably be made as a function of the researcher's stylistic preferences.

These are not the only conditions necessary to achieve model identification. When a single measured variable is posited to reflect multiple underlying factors, evaluating model identification becomes technically difficult. Bollen (1989) and Davis (1993) provided more details.

Fortunately, modern computer software is relatively sophisticated at detecting model underidentification. Most packages (e.g., AMOS) provide diagnostics such as suggestions of how many more parameters need to be "fixed" or constrained, and some guidance as to which parameters in particular are good candidates for achieving model identification.

Matrix of Association Coefficients to Analyze

As noted in chapter 3, in EFA there are numerous choices that can be made regarding the matrix of association coefficients to be analyzed. Popular EFA choices include the Pearson product–moment r matrix, the Spearman's rho matrix, and the variance/covariance matrix. As noted in chapter 3, the default choice in statistical packages is the Pearson r matrix, and this is the selection used in the vast preponderance of published EFA studies.

The statistical software for CFA supports an even larger, dizzying array of choices. These include coefficients with which many readers may be unfamiliar, such as the polyserial or the polychoric correlation coefficients.

But the default choice in most CFA software is the covariance matrix. This matrix has covariances off the main diagonal. Remember that $r_{XY} = COV_{XY} / [(SD_X)(SD_Y)]$, and that therefore $COV_{XY} = r_{XY}(SD_X)(SD_Y)$. The correlation is a "standardized" covariance from which the two standard deviations have been removed by division.

The variance/covariance matrix has variances on the main diagonal. Because the covariance of a variable with itself equals the variance of the variable, it is more common to hear researchers refer to this matrix more simply as the covariance matrix.

The computation of both the Pearson r and the covariance matrices presumes that the data are intervally scaled. This means that the researcher believes that the underlying measurement tool yielding the scores has equal units throughout. For example, on a ruler every inch (e.g., from 1″ to 2″, from 11″ to 12″) is the same amount of distance (or would be if the ruler were perfect). Note that the assumption is that the ruler has equal intervals, and *not* that the scores of different people are all equidistant.

In psychology sometimes the interval scaling of the measurement is more debatable when scores on abstract constructs are being estimated. For example, if Likert scales are used to measure attitudes, and 1 = *never*, 2 = *often*, 3 = *very often*, and 4 = *always*, most researchers would doubt that the data were intervally (continuously) scaled. However, if 1 = *strongly disagree*, 2 = *disagree*, 3 = *agree*, and 4 = *strongly agree*, many (not all) researchers would deem the scores intervally scaled, or at least approximately intervally scaled.

Of course, whenever there is some doubt regarding the scaling of data, or regarding the selection of matrix of associations to analyze, it is thoughtful practice to use several reasonable choices reflecting different premises. When factors are invariant across analytic decisions, the researcher can vest greater confidence in a view that results are not methodological artifacts.

In CFA, for some complex statistical reasons, it is almost always preferable to use the covariance matrix, rather than the Pearson *r* matrix, as the matrix to be analyzed. There are some conditions in which the Pearson product–moment matrix may be correctly used (see Cudeck, 1989). However, the covariance matrix will always yield correct results, as long as the data are appropriately scaled and distributed.

Multivariate Normality

As noted in chapter 7, the shapes of the data distributions being correlated affect the matrices of association actually analyzed in EFA. Traditionally, considerable attention is paid to these issues in Q-technique EFA applications, while less attention is focused on these concerns in R-technique EFA applications, even when the Pearson product–moment *r* matrix is the basis for analysis.

However, many CFA results are particularly affected by data distribution shapes. Certain statistical estimation theories presume multivariate normality, especially with regard to the standard errors for the estimates of model parameters. And some CFA (and SEM) model fit statistics also presume normality.

Even *univariate normality* is not as straightforward as many researchers erroneously believe. Misconceptions regarding univariate normality occur, in part, because many textbooks only present a single picture of a univariate normal distribution: z scores that are normally distributed (i.e., the "standard normal" distribution).

There is only one univariate normal distribution for a given mean and a given standard deviation. But for every different mean or every different standard deviation, the appearance of the normal distribution changes. Some normal distributions when plotted in a histogram have the classic bell shape that many associate with the normal or Gaussian curve.

But bells come in infinitely many shapes. Some bells appear fairly flat and wide, whereas others appear tall and narrow. Of course, one universally true statement about normal distributions with different appearances is that they all are unimodal and symmetrical (i.e., coefficient of skewness = 0) and have a coefficient of kurtosis of zero.

Bivariate normality is even more complicated. Because there are now scores on two variables for each person, a three-dimensional space is required to represent the distribution, and three-dimensional objects must be located in this space to represent the scores of different people. The objects used might be little BBs or ballbearings, one for each person.

An X-axis and a perpendicular Y-axis might be drawn on a piece of paper. A vertical axis extending through the X,Y intercept might be used to measure the frequencies at different score pairs. For example, if Colleen and Deborah both had scores of 1,1, their BBs would be stacked on top of each other at this particular score pair coordinate.

If both variables are z scores and normally distributed, the BBs in bivariate data would create a literal three-dimensional bell shape. If we traced around the outer edges of the BBs on the paper, a circle would be drawn on the paper. If we made a slice through the BBs anywhere up the frequency axis, parallel to but moving away from the piece of paper, concentric and narrowing circles would be defined until we reached the top of the bell. If we sliced through the BBs with a knife, in any location only with the restriction that the knife cut all the way down through the vertical frequency axis, every slice would yield a univariate standard normal distribution.

If the Y scores were more spread out than the X scores, the signature or foot of the bell defined by tracing on the paper around the BBs would be an ellipse rather than a circle. If we sliced upward on the vertical axis, but parallel to the paper, every slice would yield a concentric ellipse until we reached the top of the bell. If we sliced through the BBs with a knife, in any location only with the restriction that the knife cut all the way down through the vertical frequency axis, every slice would yield a univariate normal distribution, albeit potentially slices that were narrower or wider than each other.

Univariate normality of both variables is a necessary but not sufficient condition for bivariate normality. Bivariate normality of all possible pairs of variables is a necessary but not sufficient condition for multivariate normality. Of course, multivariate normality requires that more than three dimensions are conceptualized. These are not easy dynamics to picture or to understand. Henson (1999) provided a very useful discussion of these concepts.

There are primarily two ways that researchers can test the multivariate normality of their data. First, if more than 25 or so measured variables are involved, the graphical technique (not involving a statistical significance test) described by Thompson (1990a) can be used.

Second, a statistical test of the multivariate coefficient of kurtosis, using the null hypothesis that kurtosis equals zero, is available in commonly used CFA/SEM computer programs. These computer programs also provide a "critical ratio" (t or Wald statistic) of this parameter estimate to its standard error. Researchers hope that the multivariate kurtosis coefficient is close to zero, and that the null hypothesis is not rejected.

What can be done if the assumptions of multivariate normality cannot be met? One choice is to use a parameter estimation theory, described in a subsequent section of this chapter, that does not require this assumption. However, "distribution-free" estimation theory does require even larger sample sizes than do more commonly used theories.

A second alternative is to create item "parcels" by bundling together measured variables. For example, if eight measured variables are presumed to measure Factor I, the data can be reexpressed as four measured variables. For example, each person's score on the measured variable that is most skewed left is added to the person's score on the measured variable that is most skewed right, the score on the measured variable that is second-most skewed left is added to the score on the measured variable that is second-most skewed right, and so forth. However, this strategy is subject to the general limitations of parceling methods (cf. Bandalos & Finney, 2001; Nasser & Wisenbaker, 2003).

Outlier Screening

It is well-known that *outliers* can grossly influence statistical results for a data set, even if the number of outliers is small and the sample size is large. Outliers can influence CFA results (see Bollen, 1987) just as they can other analyses.

Yet misconceptions regarding outliers abound. Some researchers incorrectly believe that outliers have anomalous scores on all possible variables with respect to all possible statistics. One can almost visualize evil, aberrant people whose foreheads are (or should be) branded with large capital Os.

Outliers are persons with aberrant or anomalous scores on one or more variables with respect to one or more statistics. But most outliers are not outliers on all variables or with respect to all statistics. Consider the scores of Tom, Dick, and Harry on the variables X and Y:

Person	X	Y
Tom	1	2
Dick	2	4
Harry	499	998

TABLE 10.4
Deviation Scores (*x*) for Participant #166

Variable	X	M	x
PER1	9	7.65	1.35
PER2	5	7.31	−2.31
PER3	9	7.20	1.80
PER4	8	7.46	0.54
PER5	9	5.69	3.31
PER6	9	5.72	3.28
PER7	9	5.79	3.21
PER8	3	6.07	−3.07

For these data, $\overline{X} = 167.33$, $Mdn_X = 2.00$, $\overline{Y} = 334.67$, and $Mdn_Y = 4.00$. Clearly, for these data Harry is an outlier with respect to the means. But the r^2 is +1.00! Harry is not an outlier with respect to the Pearson product–moment correlation coefficient. Indeed, remove Harry from the data set and r^2 remains completely unaltered.

Notwithstanding the limitations of examining the deviation of people's scores from the means as a way of detecting outliers, a common multivariate method for identifying outliers examines exactly these deviations. The statistic reported by most CFA/SEM computer programs is called the *Mahalanobis distance* (D^2_i).

D^2_i is computed as a function of the deviations of the *i*th person's scores from the means on all the measured variables. If there were eight measured variables (e.g., PER1 through PER8 in Appendix A), the formula would be:

$$D^2_i = (\mathbf{x}_{i\,8 \times 1} - \overline{\mathbf{X}}_{8 \times 1})'_{1 \times 8} \, \mathbf{S}_{8 \times 8}^{-1} \, (\mathbf{x}_{i\,8 \times 1} - \overline{\mathbf{X}}_{8 \times 1})_{8 \times 1},$$

where $\mathbf{S}_{8 \times 8}$ is the covariance matrix for the eight measured variables.

Logically, the covariances must be considered in computing how far a person's score vector is from the means. If a person had deviation scores $(x_i = X_i - \overline{X})$ of 2.0 on two variables, these deviations should not be counted twice when both variables are perfectly correlated. Also, the spreadoutness of the variables must be taken into account. A deviation of 2.0 when $SD_X = 2$ does not have the same meaning as a deviation score of 2.0 when $SD_X = 10$.

Table 10.4 presents the deviation scores of the $i = 166$th person in the Appendix A data. The covariance matrix for these eight measured variables for an analysis limited to the 100 faculty members is presented in Table 10.5.

For these data the D_{166}^2 for the faculty member whose scores are presented in Table 10.4 equals 39.32. Most CFA/SEM computer programs provide these Mahalanobis distances as an optional output statistic. The

TABLE 10.5
Covariance and Correlation Matrices for Eight Measured Variables
(n = 100 Faculty)

Variable	Measured variable							
	PER1	PER2	PER3	PER4	PER5	PER6	PER7	PER8
PER1	1.847	*0.844*	*0.653*	*0.825*	*0.328*	*0.349*	*0.388*	*0.461*
PER2	1.768	2.374	*0.549*	*0.774*	*0.256*	*0.255*	*0.289*	*0.444*
PER3	1.340	1.278	2.280	*0.657*	*0.318*	*0.335*	*0.400*	*0.363*
PER4	1.681	1.787	1.488	2.248	*0.336*	*0.335*	*0.344*	*0.416*
PER5	0.921	0.816	0.992	1.043	4.274	*0.844*	*0.872*	*0.749*
PER6	0.962	0.797	1.026	1.019	3.543	4.122	*0.850*	*0.723*
PER7	0.986	0.835	1.132	0.967	3.375	3.231	3.506	*0.730*
PER8	1.344	1.468	1.176	1.338	3.322	3.150	2.935	4.605

Note. Covariances are not italicized; Pearson rs (i.e., standardized covariances) are italicized. Pearson r = $COV_{XY} / (SD_X \times SD_Y)$ (e.g., $0.844 = 1.768 / (1.847^{.5} \times 2.374^{.5})$). $COV_{XY} = r_{XY} \times SD_X \times SD_Y$ (e.g., $1.768 = 0.844 \times 1.847^{.5} \times 2.374^{.5}$).

distances can also be obtained within SPSS by performing a regression listing the measured variables as the predictor variables in a multiple regression analysis and using *any* other variable as the dependent variable. For this example, if participant ID number was used as the pseudo-dependent variable, the SPSS syntax would be:

```
regression variables=id per1 to per8/
  descriptive=mean stddev corr/
  dependent=id/enter per1 to per8/
  save=mahal(mahal) .
sort cases by mahal(D) .
print formats mahal(F8.4) .
list variables=id per1 to per8 mahal/cases=999 .
```

However, even when we know that a given person is an outlier on a given set of variables in regard to particular statistics, it is not always clear what should be done with the person's scores. If the researcher interviews the outlier and determines that bizarre scores were intentionally generated, then deleting the person's scores from the data set seems only reasonable. But when we cannot determine the reason for the score aberrance, it seems less reasonable to delete scores only because we are unhappy about their atypicality with respect to deviations from the means.

One resolution refuses to represent all analytic choices as being either/or. The analysis might be conducted *both* with the outlier candidates in the data and with them omitted. If interpretations are robust across various decisions, at least the researcher can be confident that the results are not artifacts of methodological choices.

It is also critical to take into account the analytic context when making these decisions. The context is very different if one is doing research in a previously unexplored area versus an arena in which numerous studies have been conducted. When other studies have been done the researcher can feel more comfortable that outliers are not compromising generalizability, if results replicate across studies. By the same token, researchers may be somewhat less concerned about outliers when they are also conducting internal replicability analyses.

As Cohen, Cohen, West, and Aiken (2003) argued,

> The extent of scrutiny of these statistics depends on the nature of the study. If the study is one of a series of replications, inspection of graphical displays for any obvious outliers is normally sufficient. *Consistency of the findings across replications provides assurance that the presence of an outlier is not responsible for the results* [italics added]. On the other hand, if the data set is unique and unlikely to be replicated (e.g., a study of 40 individuals with a rare medical disorder), very careful scrutiny of the data is in order. (pp. 410–411)

Parameter Estimation Theory

Conventional graduate courses in applied statistics teach analyses such as t tests, analysis of variance, multiple regression, descriptive discriminant analysis, and canonical correlation analysis. All of these analyses use sample data to compute sample statistics (e.g., sample means, correlation coefficients) that are then taken as estimates or corresponding population parameters (e.g., μ, σ, ρ).

All of these analyses invoke a single statistical theory for parameter estimation called *ordinary least squares* (OLS). These estimates seek to maximize the sum-of-squares$_{\text{MODEL}}$, or its multivariate equivalent, and the corresponding r^2 effect size analog (e.g., R^2, η^2) in the *sample* data.

However, there are (many) other parameter estimation theories that can also be used. Indeed, it can be argued that classical statistical analyses are not very useful, and may even routinely lead to incorrect research conclusions, because classical methods are so heavily influenced by outliers (see Wilcox, 1998).

Two alternative parameter estimation theories are fairly commonly used in CFA (and SEM). Hayduk (1987) and Jöreskog and Sörbom (1993) provided more detail on these and other choices. When sample size is large, and if the model is correctly specified, the various estimation theories tend to yield consistent and reasonably accurate estimates of population parameters (e.g., population factor pattern coefficients).

First, *maximum likelihood* theory is the default parameter estimation theory in most statistical packages. This estimation theory attempts to estimate the true population parameters, and not to reproduce the sample data or minimize the sample sum-of-squares$_{\text{UNEXPLAINED}}$ while maximizing the related sample effect size (e.g., R^2, η^2). This is accomplished by applying maximum likelihood theory to the sample covariance matrix to estimate the population covariance matrix, $\Sigma_{V \times V}$, and then deriving factors to reproduce that matrix rather than the sample covariance matrix.

Obviously, estimating population parameters with a statistical theory that optimizes this estimation (rather than features of the sample data) is very appealing. Better estimation of population parameters should yield more replicable results. This estimation theory also has the appeal that most of the statistics that evaluate model fit are appropriate to use when this theory has been invoked. However, maximum likelihood estimation theory does require that the data have at least an approximately multivariate normal distribution.

Second, *asymptotically distribution-free* (ADF) estimation theory may be invoked. As the name of this theory implies, these estimates tend to be accurate parameter estimates regardless of distribution shapes as the sample size is increasingly large. This is an appealing feature of the theory. However, very large sample sizes of more than 1,000 cases are required to invoke this theory for some analytic problems.

Of course, even CFA maximum likelihood estimation often requires larger sample sizes than would EFA modeling for the same data set (MacCallum, Browne, & Sugawara, 1996). West, Finch, and Curran (1995) reviewed some of the relevant issues and choices regarding the distributional assumptions of various parameter estimation theories.

Jackson (2001) found that sample size per se, and not the number of parameters being estimated, is the critical ingredient in deriving accurate CFA estimates. And sample size is less of a concern when the average pattern coefficient is |.80| or greater.

POSTANALYSIS DECISIONS

Once the analysis is in hand, various postanalysis decisions arise. First, the goodness of fit for each rival model must be evaluated. Second, specific errors in model specification must be considered.

Model Fit Statistics

A unique feature of CFA is the capacity to quantify the degree of model fit to the data. However, there are dozens of fit statistics, and the properties of the fit statistics are not completely understood. As Arbuckle and Wothke (1999) observed, "Model evaluation is one of the most unsettled and difficult issues connected with structural modeling [and CFA]. . . . Dozens of statistics . . . have been proposed as measures of the merit of a model" (p. 395).

Classically, Monte Carlo or simulation studies are used to explore the behaviors of statistics under various conditions (e.g., different sample sizes, various degrees of violations of distributional assumptions). In Monte Carlo research, populations with known parameters are created (e.g., with an exactly known degree of deviation from normality), and numerous random samples (e.g., 1,000, 5,000) from these populations are drawn to evaluate how the statistics perform.

Many of the previous simulation studies of CFA/SEM model fit statistics specified populations associated with known models, and then drew samples and tested the fit of the known models. Such simulations studies are *not* realistic as regards real research situations, because in real research the investigator does not usually know the true model. Indeed, if we knew the true model, we would not be doing the modeling study!

Better simulation studies of fit statistics have created populations for certain models, and then intentionally evaluated the behavior of fit statistics for various conditions and various model misspecifications. Recent simulation studies in this venue have provided important insights regarding how the fit statistics perform (cf. Hu & Bentler, 1999; Fan, Thompson, & Wang, 1999; Fan, Wang, & Thompson, 1997), but there is still a lot to learn (Bentler, 1994).

Four fit statistics are most commonly used today. In general, it is best to consult several fit statistics when evaluating model fit and to evaluate whether the results corroborate each other.

First, the χ^2 *statistical significance test* has been applied since the origins of CFA. As noted in chapter 2, in EFA the reproduced correlation matrix $(\mathbf{R}_{V \times V}{}^+)$ can be computed by multiplying $\mathbf{P}_{V \times F}$ by $\mathbf{P}_{F \times V}{}'$. The residual correlation matrix $(\mathbf{R}_{V \times V}{}^-)$ is computed by subtracting $\mathbf{R}_{V \times V}{}^+$ from $\mathbf{R}_{V \times X}$.

The same basic idea can be applied in CFA. One can estimate how much of a covariance matrix is reproduced by the parameter estimates, and how much is not. In CFA, the null hypothesis that is tested is essentially whether the residual covariance estimate equals a matrix consisting only of zeroes.

Unlike most substantive applications of null hypothesis statistical significance testing, in CFA one does *not* want to reject this null hypothesis (i.e., does *not* want statistical significance), at least for the preferred model. The degrees of freedom for this test statistic are a function of the number of measured variables and of the number of estimated parameters. For example, if there are eight measured variables, the df_{TOTAL} are 36 ([8 / (8 + 1)] / 2). If eight factor pattern coefficients, eight error variances, and one factor covariance are estimated, the degrees-of-freedom$_{MODEL}$ equal 19 (i.e., 36 − 17).

However, as with conventional statistical significance tests, though the sample size is not used in computing the degrees of freedom for the test, the sample size *is* used in computing the χ^2. Even for the same input covariance matrix and the same degree of model fit, χ^2 is larger for a larger sample size. This makes this fit statistic not very useful when sample size is large, as it usually should be in CFA, because with large sample sizes all models will tend to be rejected as not fitting. As Bentler and Bonnett (1980) observed, "in very large samples virtually all models that one might consider would have to be rejected as untenable. . . . This procedure cannot generally be justified, since the chi-square value . . . can be made small by simply reducing sample size" (p. 591).

The χ^2 test is not very useful in evaluating the fit of a single model. But the test is very useful in evaluating the comparative fits of nested models. Models are nested when a second model is the same as the first model, except that one or more additional parameters (e.g., factor covariances) are estimated or freed in the second model.

The χ^2s and their degrees of freedom are additive. For example, if for the first model positing two uncorrelated factors the χ^2 is 49.01 ($df = 20$), and for the second model estimating the same parameters except that the factor covariances are freed to be estimated the χ^2 is 31.36 ($df = 19$), then the differential fit of the two models can be statistically tested using $\chi^2 = 17.65$ (i.e., 49.01 − 31.36) with $df = 1$ (20 − 19). The $p_{CALCULATED}$ of this difference can be determined using the Excel spreadsheet function

```
=chidist(17.65,1),
```

which here yields $p = .000027$.

When maximum likelihood estimation theory is used, and the multivariate normality assumption is not perfectly met, the χ^2 statistic will be biased. However, Satorra and Bentler (1994) developed a correction for this bias that may be invoked in such cases.

Second, the *normed fit index* (NFI; Bentler & Bonnett, 1980) compares the χ^2 for the tested model against the χ^2 for the baseline model presuming that the measured variables are completely independent. For example, if the χ^2 for the tested CFA model was 31.36, and the χ^2 for the baseline independence model presuming the measured variables were perfectly uncorrelated was 724.51, the NFI would be (724.51 − 31.37) / 724.51 = 693.15 / 724.51 = 0.96. The NFI can be as high as 1.0. Generally, models with NFIs of .95 or more may be deemed to fit reasonably well.

Third, the *comparative fit index* (CFI; Bentler, 1990), like the NFI, assesses model fit relative to a baseline null or independence model. The CFI makes use of what is called the noncentral χ^2 distribution; Cumming and Finch (2001) provided an explanation of noncentral test distributions. Again, values approaching 1.0 are desired, with statistics around .95 indicating reasonable model fit.

Fourth, the *root-mean-square error of approximation* (RMSEA; Steiger & Lind, 1980) estimates how well the model parameters will do at reproducing the population covariances. A model estimated to reproduce exactly the population covariances would have an RMSEA of zero. Unlike the NFI and the CFI, we want the RMSEA to be small. Values of roughly .06 or less are generally taken to indicate reasonable model fit.

Fit statistics can also be weighted to reward model parsimony (i.e., models that spend fewer degrees of freedom to estimate parameters). Such statistics, termed *parsimony-adjusted* (Mulaik et al., 1989), are computed by multiplying fit statistics (e.g., NFI, CFI) by the ratio of the model *df* to the degrees of freedom for the baseline model.

Model Misspecification Statistics

When a model is specified and tested, there are numerous opportunities for model misspecification (i.e., model specification errors). Measured variables are used in the model that should not be. Measured variables are omitted from the model that should have been measured. The wrong analysis (e.g., estimation theory) is used. Parameters are estimated that should not be freed to be estimated. Parameters are fixed and not estimated that should have been freed.

Diagnostics for using the wrong variables are difficult to imagine. In particular, the statistical analysis cannot be expected to tell the researcher what unused variables not part of the analysis should, instead, have been used. But CFA/SEM programs do yield diagnostics regarding (a) what parameters were freed but possibly should not have been estimated and (b) what parameters were fixed and not estimated but possibly should be estimated.

Parameters Erroneously Freed

As noted in chapter 9, in various statistical analyses, we can divide a parameter estimate by its standard error to compute the critical ratio, Wald statistic, or *t* statistic, for the estimate. We usually expect these *t* values to be greater than roughly |2.0| for parameters that we wish to interpret as being nonzero. These statistics are provided routinely by CFA/SEM computer packages. Freed parameters with associated *t* statistics less than roughly |2.0| are candidates to be fixed (i.e., not estimated).

When maximum likelihood estimation theory is used, and the multivariate normality assumption is not perfectly met, the standard errors for the parameter estimates will also be inaccurate. Satorra and Bentler (1994) proposed methods for estimating the correct standard errors, and consequently correct critical ratios, using these standard errors as denominators, by taking into account data distribution shapes. Fouladi (2000) found that these corrections work reasonably well if sample sizes are 250 or more. Most modern software provides these corrections as they are needed.

Parameters Erroneously Fixed

For each parameter fixed to not be estimated that could be estimated, the CFA/SEM software can estimate the decrease (improvement) in the model fit χ^2 resulting from freeing a given previously fixed parameter instead to be estimated. In some packages (e.g., AMOS) these *modification indices* are "lower-bound" or conservative estimates, and if a fixed parameter is freed the χ^2 reduction predicted by the software may be better than predicted. Obviously, fixed parameters with the largest modification indices are among the leading candidates for model respecification.

Model Specification Searches

Confirmatory factor analysis is intended to be, as the name implies, confirmatory. Respecifying a CFA model based on consultation of critical ratio and modification index statistics is a dicey business, *if* the same sample is being used to generate these statistics and then to test the fit of the respecified model.

Using the same sample to respecify the model, and then test the respecified model, increases capitalization on sampling error variance and decreases the replicability of results. And using the same sample in this manner also turns the analysis into an explanatory one, albeit using CFA algorithms.

As Thompson (2000c) emphasized, model respecification should *never* "be based on blind, dust-bowl empiricism. Models should only be respecified in those cases in which the researcher can articulate a persuasive rationale as to why the modification is theoretically and practically defensible" (p. 272). Model respecification is reasonable when each change can be justified theoretically, and when the fit of the respecified model can be evaluated in a holdout or independent sample.

MAJOR CONCEPTS

Exploratory factor analysis and CFA are part of the same general linear model (Bagozzi et al., 1981). However, CFA models can estimate some parameters that cannot be evaluated within EFA. For example, only in CFA can the covariances of error variances be nonzero and estimated. In EFA factors can be either correlated or uncorrelated, but in CFA additionally some factors may be allowed to be correlated whereas other pairs of factors in the same model are constrained to be uncorrelated. Or, uniquely in CFA, factor correlations can be freed but constrained to be equal. And in CFA, unlike EFA, all the pattern coefficients for a factor can be freed to be estimated but constrained to be equal.

Confirmatory factor analysis models require the specification of one or more hypothesized factor structures, and the analyses directly test and quantify the degrees of fit of rival models. Because more than one model may fit in a given study, it is important to test the fit of multiple plausible models, so that any case for fit of a given model is more persuasive.

Only "identified" factor models can be tested in CFA. In CFA, degrees of freedom are a function of number of measured variables and not of sample size. A necessary but not sufficient condition for model identification is that the number of estimated parameters (e.g., pattern coefficients, factor correlations, error variances) cannot exceed the available degrees of freedom. Ideally, considerably fewer parameters will be estimated than there are degrees of freedom, so that the model is more disconfirmable. Fit when more degrees of freedom are unspent is inherently more impressive. Some factor pattern coefficients or factor variances must also be constrained to obtain model identification.

Like EFA, CFA requires various decisions, such as what matrix of associations (e.g., covariance matrix) to analyze, and what estimation theory (e.g., maximum likelihood) to use. Some estimation theories require that data are multivariate normal. And screening data for outliers can be particularly important when very elegant methods, such as CFA, are brought to bear.

Confirmatory factor analysis model tests yield various statistics that quantify degree of model fit. It is generally best practice to consult an array of fit statistics (Fan et al., 1999; Hu & Bentler, 1999) in the hope that they will confirm each other.

The analysis also yields critical ratios that evaluate whether freed parameters were indeed necessary, and modification indices that indicate whether fixed parameters instead should have been estimated. Respecifying models based on such results is a somewhat dicey proposition, unless theory can be cited to justify model revisions. Such "specification searches" are, however, completely reasonable and appropriate when the respecified model is tested with a new, independent sample.

11

SOME CONFIRMATORY FACTOR ANALYSIS INTERPRETATION PRINCIPLES

Many principles of confirmatory factor analysis (CFA) mirror those in exploratory factor analysis (EFA; e.g., the importance of consulting *both* pattern and structure coefficients), given the relationships of these two analyses within the general linear model (GLM). However, there are also aspects of CFA that are unique to CFA model tests.

For example, in EFA we almost always use only standardized factor pattern coefficients. However, in CFA *both* unstandardized pattern coefficients (analogous to regression unstandardized *b* weights) and standardized pattern coefficients (analogous to regression standardized β weights) are computed.

The statistical significance tests of individual CFA pattern coefficients are conducted using the unstandardized coefficients. These coefficients are computed by dividing a given parameter by its respective standard error to compute a *t* (or Wald or critical ratio) statistic. Noteworthy unstandardized coefficients using these tests also indicate the noteworthiness of the corresponding standardized parameters, as in multiple regression.

However, when parameters are compared with each other, usually the standardized coefficients are reported, again as in multiple regression. The standardization allows apples-to-apples comparisons of related parameters within a single model.

Five aspects of CFA practice are elaborated here. These include (a) alternative ways to create identified models, (b) some estimation theory choices, (c) the importance of consulting both pattern and structure coefficients when CFA factors are correlated, (d) comparing fits of nested models, and (e) extraction of higher-order factors.

In these illustrative analyses, the Appendix A data provided by the 100 faculty members for the first eight items are used. Readers may find it useful to conduct parallel analyses using the appended data provided by the 100 graduate students.

MODEL IDENTIFICATION CHOICES

As explained in chapter 10, parameters cannot be estimated unless the model is "identified." A model is unidentified or underidentified if for that model an infinite number of estimates for the parameters are equally plausible.

A necessary but insufficient condition for identification is that the number of parameters being estimated must be equal to (just identified) or less than (overidentified) the available degrees of freedom. In this example the number of measured variables is eight. Therefore, the available degrees of freedom are 36 ($[8 \times (8 + 1)] / 2$).

The models tested here presume that items PER1 through PER4 reflect only an underlying factor named "Service Affect" and that items PER5 through PER8 reflect only an underlying factor named "Library as Place." Thus, a total of up to eight factor pattern coefficients are estimated. Also, the errors variances for each of the eight measured variables are estimated. It is assumed here that the factors are uncorrelated.

However, for this model we cannot estimate these 16 ($8 + 8$) parameters because the scales of the two latent factors have not been indicated. These scales of measurement of latent constructs are arbitrary. That is, there is no intrinsic reason to think that "Library as Place" scores in the population should have a particular variance or standard deviation. But we must declare some scale for these latent variables, or with each of infinitely many different possible variances there would be infinitely many corresponding sets of other parameters being estimated.

There are two basic ways to identify CFA models. First, for a given factor we can "fix" or constrain *any* given unstandardized factor pattern coefficient to be *any* mathematically plausible number. However, usually the number 1.0 is used for this purpose. In doing so, we are scaling the variance of the latent construct to be measured as some function of the measured variable for which we fix the pattern coefficient. Some researchers prefer to fix the pattern coefficient for the measured variable believed to be most psychometrically reliable. However, the selection is basically arbitrary.

Second, we can estimate all the pattern coefficients and fix the factor variances to any mathematically plausible or "admissible" number. Because variances cannot be negative, admissible numbers for variance estimates may not be negative. And the factor variance cannot be zero, or we would be evaluating the influences of a variable that was no longer a variable.

When factor variances are fixed, usually the variance parameters are fixed to be 1.0. This approach has a singular advantage: When we are estimating factor covariances, the covariances of the factors with each other now become factor correlations, because given that $r_{I \times II} = COV_{I \times II} / (SD_I \times SD_{II})$, when the factor variances are both 1.0, we obtain $COV_{I \times II} / (1 \times 1)$. This makes interpretation more straightforward.

In addition, t statistics are only obtained for estimated factor pattern coefficients. So it is helpful to know the statistical significance, or Wald ratios, for each factor pattern coefficient, and these would not be available if the model were identified by constraining some factor pattern coefficients.

The approach also has the advantage that all the pattern coefficients for a given factor can be compared with each other, because they were all freed. In general, we may expect that pattern coefficients on a given factor will be similar in magnitudes.

Table 11.1 compares three analyses of these data with three different model identification strategies having been invoked. In Model 1 arbitrarily the unstandardized pattern coefficient for variable PER1 on Factor I ("Service Affect") was fixed to equal 1.0, and the unstandardized pattern coefficient for variable PER8 on Factor II ("Library as Place") was fixed to equal 1.0. The factor variances were both "freed" to be estimated.

In Model 2 arbitrarily the unstandardized pattern coefficient for variable PER2 on Factor I ("Service Affect") was fixed to equal 1.0, and the unstandardized pattern coefficient for variable PER5 on Factor II ("Library as Place") was fixed to equal 1.0. The factor variances were both "freed" to be estimated.

In Model 3, portrayed graphically in Figure 11.1, all eight pattern coefficients were estimated. However, to identify the model, both factor variances were constrained to be 1.0.

Note from Table 11.1 that the standardized pattern coefficients and the factor structure coefficients are identical across the three models, regardless of how the models were identified. And the fit statistics are identical. So my preference for identifying CFA models by fixing factor variances is arbitrary but permissible as a matter of stylistic choice.

TABLE 11.1

Standardized Factor Pattern/Structure Coefficients and Fit Statistics Derived With Identification Achieved Three Ways (Maximum Likelihood; $n = 100$ Faculty)

Variable/ statistic	Model #1				Model #2				Model #3			
	Pattern		Structure		Pattern		Structure		Pattern		Structure	
	I	II	I	II	I	II	I	II	I	II	I	II
PER1	(.950)	—	0.950	0.000	0.950	—	0.950	0.000	0.950	—	0.950	0.000
PER2	0.882	—	0.882	0.000	(.882)	—	0.882	0.000	0.882	—	0.882	0.000
PER3	0.686	—	0.686	0.000	0.686	—	0.686	0.000	0.686	—	0.686	0.000
PER4	0.877	—	0.877	0.000	0.877	—	0.877	0.000	0.877	—	0.877	0.000
PER5	—	0.934	0.000	0.934	—	(.934)	0.000	0.934	—	0.934	0.000	0.934
PER6	—	0.908	0.000	0.908	—	0.908	0.000	0.908	—	0.908	0.000	0.908
PER7	—	0.933	0.000	0.933	—	0.933	0.000	0.933	—	0.933	0.000	0.933
PER8	—	(.793)	0.000	0.793	—	0.793	0.000	0.793	—	0.793	0.000	0.793
Factor variance	1.667	2.899			1.845	3.730			(1.00)	(1.00)		
χ^2	48.27				48.27				48.27			
df	20				20				20			
χ^2/df	2.41				2.41				2.41			
NFI	0.933				0.933				0.933			
CFI	0.959				0.959				0.959			
RMSEA	0.119				0.119				0.119			

Note. Parameter estimates "fixed" to be zeroes are reported as dashes ("—"). Parameter estimates "fixed" to equal 1.0 for the unstandardized model are reported in parentheses and bold. NFI = normed fit index; CFI = comparative fit index; RMSEA = root-mean-square error of approximation.

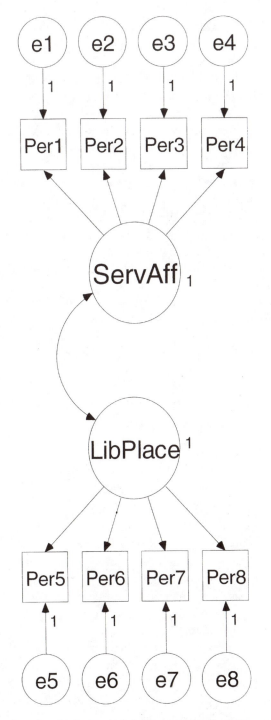

Figure 11.1. Confirmatory factor analysis factor model with eight measured variables and two correlated factors with factor variances constrained to equal 1.0. ServAff = "Service Affect"; LibPlace = "Library as Place."

DIFFERENT ESTIMATION THEORIES

As noted in chapter 10, classical statistics (e.g., analysis of variance, regression, canonical correlation analysis) invoke a statistical estimation theory called *ordinary least squares* (OLS). However, there are numerous other statistical estimation theories that can be invoked to estimate model parameters. The theory most commonly used in CFA (and in structural equation modeling [SEM]) is maximum likelihood theory. Most of the fit statistics have been developed for this theory.

However, maximum likelihood estimation does presume that data are multivariate normally distributed. Another estimation theory, asymptotically distribution-free (ADF) theory, can be invoked if the distributional assumptions of maximum likelihood cannot be met. However, ADF estimation does presume large sample size.

For comparative purposes, Table 11.2 presents the standardized pattern and the structure coefficients for the Model 3 fit to the faculty data using three different estimation theories. Note that the pattern and structure coefficients estimated in all three models are reasonably comparable.

These analyses were compared primarily for heuristic purposes. However, it is not entirely a bad thing to invoke different analytic strategies for the same data, just to confirm that results are not artifacts of analytic choices. In this instance, probably only the maximum likelihood estimates would be reported, but the researcher would sleep easier knowing that similar factors emerged for the other analyses.

CONFIRMATORY FACTOR ANALYSIS PATTERN AND STRUCTURE COEFFICIENTS

In EFA, orthogonal solutions almost always provide simple structure, are the default models in statistical packages, and dominate the literature. However, in CFA, it is more usual to test either correlated factors models or both uncorrelated and correlated factors models. That is, usually both CFA models are reasonably plausible, and we want to test plausible rival models to have the strongest possible support for our preferred model, presuming that this model ultimately has best fit.

When factors are uncorrelated, the CFA standardized factor pattern and the structure coefficients for given variables on given factors exactly equal each other, as reflected in the illustrative results reported in Table 11.2. But when CFA factors are correlated, exactly as is the case in EFA, factor pattern and factor structure coefficients are *no longer equal*.

For multiple regression (Courville & Thompson, 2001; Thompson & Borrello, 1985), descriptive discriminant analysis (Huberty, 1994), and

TABLE 11.2
Factor Pattern/Structure Coefficients and Fit Statistics for Model #3 Derived From Three Estimation Theories
(n = 100 Faculty)

Variable/ statistic	Unweighted LS Pattern I	Pattern II	Structure I	Structure II	ADF Pattern I	Pattern II	Structure I	Structure II	Maximum likelihood Pattern I	Pattern II	Structure I	Structure II
PER1	0.949	—	0.949	0.000	0.908	—	0.908	0.000	0.950	—	0.950	0.000
PER2	0.855	—	0.855	0.000	0.926	—	0.926	0.000	0.882	—	0.882	0.000
PER3	0.686	—	0.686	0.000	0.811	—	0.811	0.000	0.686	—	0.686	0.000
PER4	0.904	—	0.904	0.000	0.908	—	0.908	0.000	0.877	—	0.877	0.000
PER5	—	0.936	0.000	0.936	—	0.977	0.000	0.977	—	0.934	0.000	0.934
PER6	—	0.908	0.000	0.908	—	0.886	0.000	0.886	—	0.908	0.000	0.908
PER7	—	0.929	0.000	0.929	—	0.899	0.000	0.899	—	0.933	0.000	0.933
PER8	—	0.794	0.000	0.794	—	0.863	0.000	0.863	—	0.793	0.000	0.793
χ^2	1810.36				49.08				48.27			
df	n.a.				20				20			
χ^2/df	n.a.				2.45				2.41			
NFI	0.811				0.743				0.933			
CFI	n.a.				0.822				0.959			
RMSEA	n.a.				0.121				0.119			

Note. Parameter estimates "fixed" to be zeroes are reported as dashes ("—"). Unweighted LS = unweighted least squares; ADF = asymptotically distribution-free (ADF) theory; n.a. = not applicable; NFI = normed fit index; CFI = comparative fit index; RMSEA = root-mean-square error of approximation.

canonical correlation analysis (cf. Cohen & Cohen, 1983, p. 456; Levine, 1977, p. 20; Meredith, 1964; Thompson, 2000a), when weights and structure coefficients are unequal, sensible interpretation usually requires examination of both sets of results. Structure coefficients are equally important in CFA (cf. Graham et al., 2003; Thompson, 1997). Thompson, Cook, and Heath (2003) presented illustrative strategies for reporting results for correlated CFA factors.

Put differently, as Bentler and Yuan (2000) emphasized, in CFA when factors are correlated, "*even if* a factor does not influence a variable, that is, its pattern coefficient or weight is zero, the corresponding structure coefficient, representing the correlation or covariance of the variable with that factor, generally will *not* be zero" (p. 327). There are three possible situations that arise with correlated factors, not counting combinations of these situations.

Univocal Models

One possibility is that each measured variable reflects the influences of only a single one of the correlated factors. Model 4 presented in Table 11.3 involves such a case.

For such "univocal" models in which each measured variable speaks on behalf of a single latent variable, the freed standardized parameters and the factor structure coefficients always exactly match. For example, PER1 had a pattern coefficient of Factor I in Model 4 of .950; the corresponding structure coefficient also was exactly .950.

However, note that *none* of the structure coefficients associated with pattern coefficients fixed to equal zeroes were themselves zeroes. For example, PER1 had a pattern coefficient on Factor II in Model 4 fixed to equal zero, but the corresponding structure coefficient was .398.

In other words, when factors are correlated, all measured variables are correlated with *all* factors. This dynamic must be considered as part of result interpretation. The higher the factor correlations are in absolute values, the more all variables will be correlated with all the factors.

Correlated Error Variances

A unique feature of CFA versus EFA is the possibility of estimating covariances or correlations of measurement error variances. If the model is correctly specified, then the error variances of the measured variables reflect score unreliability (see Thompson, 2003).

For example, the variance of the measured variable PER1 was 1.847. In Model 4 the estimated error variance for this variable was .178. This means that, if the model is correctly specified, the unreliability of this

TABLE 11.3
Parameter Estimates and Fit Statistics for Three Models Positing Correlated Factors ($n = 100$ Faculty)

Variable/ statistic	Model #4 Pattern I	II	Structure I	II	Model #5 Pattern I	II	Structure I	II	Model #6 Pattern I	II	Structure I	II
PER1	0.950	—	0.950	0.398	0.938	—	0.938	0.393	0.949	—	0.949	0.378
PER2	0.879	—	0.879	0.368	0.890	—	0.890	0.373	0.882	—	0.882	0.352
PER3	0.690	—	0.690	0.289	0.716	—	0.716	0.300	0.688	—	0.688	0.274
PER4	0.877	—	0.877	0.367	0.881	—	0.881	0.369	0.876	—	0.876	0.349
PER5	—	0.932	0.390	0.932	—	0.932	0.390	0.932	—	0.934	0.372	0.934
PER6	—	0.907	0.380	0.907	—	0.907	0.380	0.907	—	0.908	0.362	0.908
PER7	—	0.934	0.391	0.934	—	0.934	0.391	0.934	—	0.933	0.372	0.933
PER8	—	0.798	0.334	0.798	—	0.798	0.334	0.798	**0.206**	0.711	**0.489**	**0.793**
$COV_{e2,e3}$	n.a.				**-.277**				n.a.			
Factor r	.419				.419				.398			
χ^2	31.36				26.91				22.88			
df	19				18				18			
χ^2/df	1.65				1.50				1.27			
NFI	0.957				0.963				0.968			
CFI	0.982				0.987				0.993			
RMSEA	0.081				0.071				0.052			

Note. Parameter estimates "fixed" to be zeroes are reported as dashes ("—"). n.a. = not applicable; NFI = normed fit index; CFI = comparative fit index; RMSEA = root-mean-square error of approximation.

measured variable could be estimated as .178 / 1.847 = .096 (or 9.6%). The corresponding reliability of this measured variable could be estimated as (1.847 − .178) / 1.847 = 1.669 / 1.847 = .904 (or 90.4%; or 100% − 9.6%).

Measurement error variances in classical statistics are usually taken as being random and therefore independent. However, in some cases elements of measurement error may be systematic and overlapping. For example, in a study involving perceptions of gender roles, if two items in a given study invoked strong homophobic emotional reactions, the error variances for these two items might be correlated even though this dynamic is extraneous to the primary features of the model.

Model 5 allows the two error variances for the Factor I measured variables, "Giving users individual attention" and "Employees who deal with users in a caring fashion," to covary. This additional single parameter estimate costs one additional degree of freedom, as reported in Table 11.3. But only allowing error covariances to be nonzero does not change the structure coefficients associated with freed pattern coefficients, and the freed pattern coefficients and related structure coefficients will remain equal.

Multivocal Models

Model 6 is presented graphically in Figure 11.2. In this model there is one measured variable (PER8) through which both factors speak. Note that for such variables the freed pattern coefficients no longer equal the associated structure coefficients (i.e., .206 ≠ .489, and .711 ≠ .793).

COMPARISONS OF MODEL FITS

In any CFA (or SEM) model testing, multiple models may fit the same data. Thus, it is not best practice to test only a single, preferred model (Thompson, 2000c). Instead, it is desirable to test multiple plausible rival models. Then, if the preferred model is the only model with reasonable fit, the case for that model is correspondingly stronger.

In a factor analytic context, rival models would often include a model specifying independence of the measured variables (i.e., no factors), a model positing a single factor, and models positing both uncorrelated and correlated factors. Table 11.4 presents fit statistics for these models for these data.

The statistical significance test for the baseline independence model positing no factors results in rejection of the null hypothesis that the model fits. Note that the $p_{CALCULATED}$ value for this test is reported in scientific notation as "1.45E-134." This means that after the decimal, the p is 133 zeroes followed by "145."

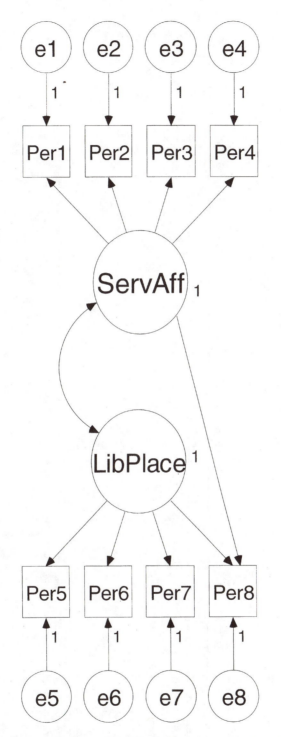

Figure 11.2. Confirmatory factor analysis Model #6 (Factors I and II multivocal through PER8). ServAff = "Service Affect"; LibPlace = "Library as Place."

TABLE 11.4
Selected Fit Statistics for Four Models and Tests of Model Differences ($n = 100$ Faculty)

Model	χ^2	df	$p_{CALCULATED}$	χ^2/df	NFI	CFI	RMSEA
Independence	724.51	28	1.45E−134	25.88	0.000	0.000	0.501
One factor	285.67	20	6.75E−49	14.28	0.606	0.619	0.366
Difference	438.845	8	9.07E−90				
Uncorrelated factors (#3)	48.27	20	0.00039	2.41	0.933	0.959	0.119
Correlated factors (#4)	31.36	19	0.03685	1.65	0.957	0.982	0.081
Difference	16.91	1	0.00004				

Note. In scientific notation, "1.45E-134" means move the decimal 134 places to the left, which would mean after the decimal there would be 133 zeroes, followed by "145." NFI = normed fit index; CFI = comparative fit index; RMSEA = root-mean-square error of approximation.

The p is *not* equal to zero, regardless of what the statistical package may print to relatively few decimal places. A common error in published literature is erroneously reporting that $p = .000$, which means that the researcher impossibly obtained an impossible result (which is impossible). Such small values can be accurately reported as $p < .001$, or exact values can be derived using Excel spreadsheet functions such as:

```
=chidist(724.51,28),
```

which here yields p = "1.45E-134."

For these tests, Model 4 provides reasonable fit. This is true notwithstanding the p value (.03685) associated with the test. As noted in chapter 10, the p values for model fit tests are inflated by the large sample sizes usually used in CFA, as are statistical significance tests in other contexts (Thompson, 1996).

The significance test is more useful in looking at *comparative fit* of nested models. In this example, Model 4 posited correlated factors, estimated 17 parameters (8 pattern coefficients, 8 error variances, and 1 covariance), and retained 19 (36 – 17) degrees of freedom. Model 3 posited uncorrelated factors, estimated 16 parameters (8 pattern coefficients, 8 error variances), and retained 20 (36 – 16) degrees of freedom.

The models are nested because Model 3 is Model 4 with one parameter (a covariance) removed. This means that the differences in the model fits can be quantified by subtracting the χ^2 values (48.27 – 31.36 = 16.91) and the corresponding degrees of freedom (20 – 19 = 1). The improvement in Model 4 over Model 3 is statistically significant ($p = .00004$).

HIGHER-ORDER MODELS

As explained in chapter 6, in EFA higher-order factors can be extracted where lower-order factors are correlated. Given the GLM, the same principles apply in CFA. One difference in the analyses, however, is that higher-order factors in CFA are extracted with the factor covariances/correlations fixed to be zeroes. In CFA higher-order analysis, the factor covariances are posited to be accounted for by the higher-order factors.

Three First-Order Factors

Figure 11.3 presents a higher-order model involving 12 measured variables and 3 first-order factors for the data provided by the 100 faculty. In CFA higher-order analyses, usually there are three or more first-order factors so that the higher-order parameters will become identified (see Byrne, 1998, pp. 170–173).

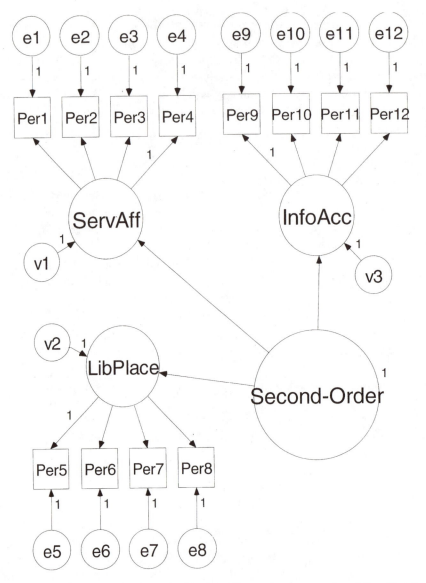

Figure 11.3. Second-order model with three first-order factors. ServAff = "Service Affect"; LibPlace = "Library as Place"; InfoAcc = "Information Access."

The model graphically portrayed in Figure 11.3 has 78 degrees of freedom ([12 × (12 + 1)] / 2). The model expends 27 of the 78 degrees of freedom to estimate 12 error variances of the measured variables, 9 first-order factor pattern coefficients, 3 variances of the first-order factors, and 3 second-order factor pattern coefficients (12 + 9 + 3 + 3 = 27). Three first-order pattern coefficients were fixed to equal 1s, and the second-order factor's variance was fixed to equal 1.

Table 11.5 presents the parameter estimates and fit statistics for this analysis. As before, note that pattern coefficients fixed to equal zero do not imply that corresponding structure or other correlations are also zero.

Note that these results suggest a reasonable model fit. These three latent variables do measure a second-order factor. But the "Information Access" first-order factor has the highest correlation with the second-order factor ($r = .990$). The first-order factor, "Library as Place," has the lowest of the three correlations with the second-order factor ($r = .602$). This analysis reflects the richness of second-order factor analysis and its capacity to model complex hierarchical dynamics.

Two First-Order Factors

A second-order CFA corresponding to Model #4 can be executed, if additional constraints are imposed so that the model becomes identified, because there are only two first-order factors. The constraint illustrated in Figure 11.4 is imposing a restriction that the two second-order pattern coefficients for the first-order factors are to be estimated, but that these estimates must be equal. In SEM/CFA software, such restrictions are communicated by using a single letter for two or more parameters constrained to be equal.

Table 11.6 presents the parameter estimates and the fit statistics associated with this model. Note that the fit statistics reported in Table 11.6 for this second-order model are identical to the fit statistics reported in Table 11.3 for the correlated first-order factor model, Model #4.

In CFA, higher-order factor analysis reexpresses the correlations among the first-order factors by modeling these dynamics in an alternative model. However, the fits of these higher-order models and the corresponding first-order models are the same, when there are fewer than four first-order factors for a given second-order factor. Nevertheless, the higher-order analysis does provide additional insight, because only in this model are the correlations of measured variables and first-order factors with higher-order constructs quantified.

MAJOR CONCEPTS

The decision of how to identify a factor model is arbitrary and does not impact fit statistics. However, stylistically, model identification via fixing factor variances to 1 has some advantages, such as making factor covariances also equal factor correlations and allowing comparisons of estimated pattern coefficients for all the measured variables on a given factor.

TABLE 11.5

Factor Pattern/Structure Coefficients and Fit Statistics for Second-Order Factor Analysis With 12 Measured Variables ($n = 100$ Faculty)

Variable/ statistic	First-order						Correlations measured by second-order
	Pattern			Structure			
	I	II	III	I	II	III	
PER1	0.944	—	—	0.944	0.395	0.650	0.656
PER2	0.882	—	—	0.882	0.369	0.607	0.613
PER3	0.692	—	—	0.692	0.290	0.477	0.481
PER4	0.882	—	—	0.882	0.369	0.608	0.614
PER5	—	0.933	—	0.390	0.933	0.556	0.561
PER6	—	0.908	—	0.380	0.908	0.541	0.546
PER7	—	0.932	—	0.390	0.932	0.555	0.561
PER8	—	0.798	—	0.334	0.798	0.475	0.480
PER9	—	—	0.576	0.397	0.343	0.576	0.571
PER10	—	—	0.716	0.493	0.426	0.716	0.709
PER11	—	—	0.663	0.456	0.395	0.663	0.656
PER12	—	—	0.480	0.331	0.286	0.480	0.476

	Second-order	
	Pattern	Structure
I	0.696	0.696
II	0.602	0.602
III	0.990	0.990
χ^2	68.94	
df	51	
χ^2/df	1.35	
NFI	0.922	
CFI	0.978	
RMSEA	0.060	

Note. Parameter estimates "fixed" to be zeroes are reported as dashes ("—"). NFI = normed fit index; CFI = comparative fit index; RMSEA = root-mean-square error of approximation.

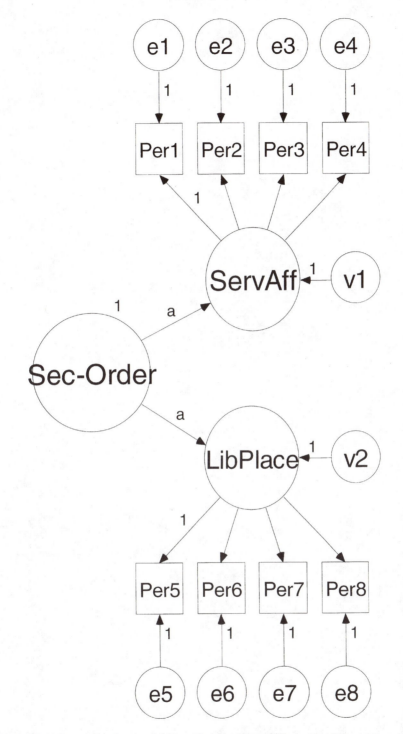

Figure 11.4. Second-order model (Sec-Order) with two first-order factors. ServAff = "Service Affect"; LibPlace = "Library as Place."

TABLE 11.6

Factor Pattern/Structure Coefficients and Fit Statistics for Second-Order Factor Analysis With Eight Measured Variables
(n = 100 Faculty)

Variable/ statistic	First-order Pattern		First-order Structure		Correlations measured by second-order
	I	II	I	II	
PER1	0.950	—	0.950	0.398	0.751
PER2	0.879	—	0.879	0.368	0.694
PER3	0.690	—	0.690	0.289	0.545
PER4	0.877	—	0.877	0.367	0.693
PER5	—	0.932	0.390	0.932	0.494
PER6	—	0.907	0.380	0.907	0.481
PER7	—	0.934	0.391	0.934	0.495
PER8	—	0.798	0.334	0.798	0.423

	Second-order Pattern	Second-order Structure
I	0.790	0.790
II	0.530	0.530
χ^2	31.36	
df	19	
χ^2/df	1.65	
NFI	0.957	
CFI	0.982	
RMSEA	0.081	

Note. Parameter estimates "fixed" to be zeroes are reported as dashes ("—"). For pattern coefficients all critical ratios > |2.0|. NFI = normed fit index; CFI = comparative fit index; RMSEA = root-mean-square error of approximation.

Just as both weights and structure coefficients are critical to correct result interpretation throughout the GLM, both pattern and structure coefficients are important in CFA when the factors are correlated (Graham et al., 2003). And just as extracting higher-order factors whenever factors are correlated is recommended, in CFA higher-order analyses provide useful insights into the structure underlying data.

12

TESTING MODEL INVARIANCE

The *sine qua non* of research is finding effects that replicate under stated conditions. When factor analytic results replicate across samples or analytic decisions, the results are said to be *invariant*. Confirmatory factor analysis (CFA) is ideally suited for testing invariance, because CFA allows constraining different types of result equalities (e.g., invariance of pattern coefficients, invariance of factor correlations) in different models and quantifies the degrees of model invariance across these variations.

Here three sets of invariance analyses are reported using the Appendix A data. The examples involve all 12 measured variables and the graduate student (n_1 = 100) and faculty (n_2 = 100) data. These analyses illustrate testing factor invariance across different respondent groups. Alternatives are equally plausible, such as investigating invariance over randomly created subgroups, across data collected from the same people over time, or across the same role group but different geographic origins (e.g., students from different regions of the country).

SAME MODEL, DIFFERENT ESTIMATES

The least restrictive test of factor invariance presumes only that the same model fits in different groups but that the parameters estimated in the different groups are independently estimated in each group. Figure 12.1 specifies an input model for such an analysis.

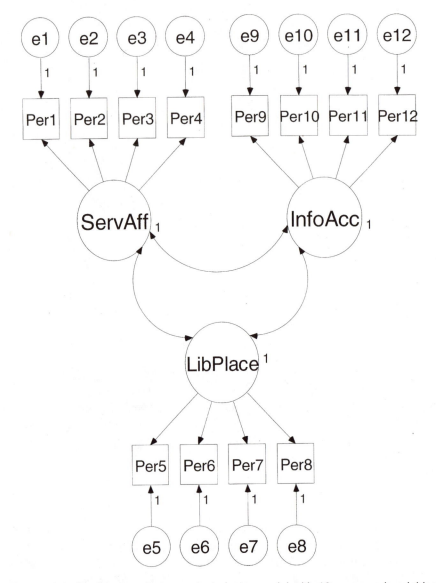

Figure 12.1. Confirmatory factor analysis factor model with 12 measured variables and 3 correlated factors with factor variances constrained to equal 1.0. ServAff = "Service Affect"; InfoAcc = "Information Access"; and LibPlace = "Library as Place."

In this analysis in both groups there are 78 degrees of freedom, because there are 12 measured variables ($[12 \times (12 + 1)] / 2$). The specified model posits that each of three correlated factors, with factor variances constrained to equal 1s, are estimated. Parameters freed to be estimated in both groups are 12 factor pattern coefficients, 12 error variances for the measured variables, and 3 factor covariances/correlations ($12 + 12 + 3 = 27$).

Thus, there are in both groups 51 (78 − 27) unspent degrees of freedom, for a total of 102 (2 × 51) available degrees of freedom after parameter estimation.

Table 12.1 presents the standardized factor pattern coefficients, the factor structure coefficients, the factor covariances (here also correlations, because the factor variances are all fixed to equal 1), and fit statistics for the analysis. The fit statistics suggest that the model is a plausible fit to the data provided by both groups.

There are some differences in the parameters estimated independently across the two groups. For example, the item PER3 ("Employees who deal with users in a caring fashion") is more correlated with the "Service Affect" factor for graduate students ($r_S^2 = .835^2 = 69.7\%$) than for faculty ($r_S^2 = .692^2 = 47.9\%$). The difference in the squared structure coefficients, which can be directly compared because squaring has been invoked, is noteworthy. Students define affect of service in libraries more than faculty in terms of being dealt with in a caring fashion by library staff.

Conversely, the item PER6 ("A meditative place") is more correlated with the "Library as Place" factor for faculty ($r_S^2 = .908^2 = 82.4\%$) than for students ($r_S^2 = .753^2 = 56.7\%$). In this case, faculty perceptions of the library are more correlated with a quiet and contemplative atmosphere.

The factor correlations are reasonably similar. The biggest differences occur on the correlation of "Service Affect" with "Library as Place" for graduate students ($r_{I \times II}^2 = .569^2 = 32.4\%$) versus faculty ($r_{I \times II}^2 = .418^2 = 17.5\%$).

Nevertheless, the single model fit reasonably well in both groups, and the parameter estimates were reasonably similar across groups. Furthermore, the detected differences seem reasonably to reflect natural differences in needs and interests of the two groups of library users.

SAME MODEL, EQUAL ESTIMATES

Figure 12.2 presents a more constrained test of factor invariance across the two groups. In this case the three unstandardized factor pattern coefficients are constrained to equal 1s. The remaining nine pattern coefficients are freed to be estimated, but with the constraint that *the set of nine understandardized estimates must be equal across the two groups,* as reflected by the use of letters within the drawing of the model. Also estimated independently in the two groups are 12 error variances for the measured variables, the variances of the 3 factors, and the 3 factor covariances.

Thus, from the available total of 156 degrees of freedom (78 × 2), a total of 45 (9 + 2 [12 + 3 + 3] = 9 + 2 [18] = 9 + 36) degrees of freedom are spent on parameter estimation, leaving 111 degrees of freedom unspent.

TABLE 12.1
Factor Pattern/Structure Coefficients and Fit Statistics for
First-Order Factor Analysis With 12 Measured Variables
(n_1 = 100 Graduate Students; n_2 = 100 Faculty)

Group/ variable/ statistic	Pattern			Structure		
	I	II	III	I	II	III
Graduate students						
PER1	0.945	—	—	0.945	0.538	0.607
PER2	0.886	—	—	0.886	0.504	0.569
PER3	0.835	—	—	0.835	0.475	0.536
PER4	0.822	—	—	0.822	0.468	0.528
PER5	—	0.907	—	0.516	0.907	0.514
PER6	—	0.753	—	0.429	0.753	0.427
PER7	—	0.890	—	0.507	0.890	0.505
PER8	—	0.834	—	0.475	0.834	0.473
PER9	—	—	0.827	0.531	0.469	0.827
PER10	—	—	0.739	0.474	0.419	0.739
PER11	—	—	0.813	0.522	0.461	0.813
PER12	—	—	0.613	0.393	0.347	0.613
$r_{I \times II}$	0.569					
$r_{I \times III}$	0.642					
$r_{II \times III}$	0.567					
Faculty						
PER1	0.944	—	—	0.944	0.395	0.650
PER2	0.882	—	—	0.882	0.369	0.607
PER3	0.692	—	—	0.692	0.290	0.477
PER4	0.882	—	—	0.882	0.369	0.608
PER5	—	0.933	—	0.390	0.933	0.556
PER6	—	0.908	—	0.380	0.908	0.541
PER7	—	0.932	—	0.390	0.932	0.555
PER8	—	0.798	—	0.334	0.798	0.475
PER9	—	—	0.576	0.397	0.343	0.576
PER10	—	—	0.716	0.493	0.426	0.716
PER11	—	—	0.663	0.456	0.395	0.663
PER12	—	—	0.480	0.331	0.286	0.480
$r_{I \times II}$	0.418					
$r_{I \times III}$	0.689					
$r_{II \times III}$	0.596					
χ^2	135.03					
df	102					
χ^2/df	1.32					
NFI	0.924					
CFI	0.980					
RMSEA	0.040					

Note. Parameter estimates "fixed" to be zeroes are reported as dashes ("—"). All critical ratios > |2.0|.
NFI = normed fit index; CFI = comparative fit index; RMSEA = root-mean-square error of approximation.

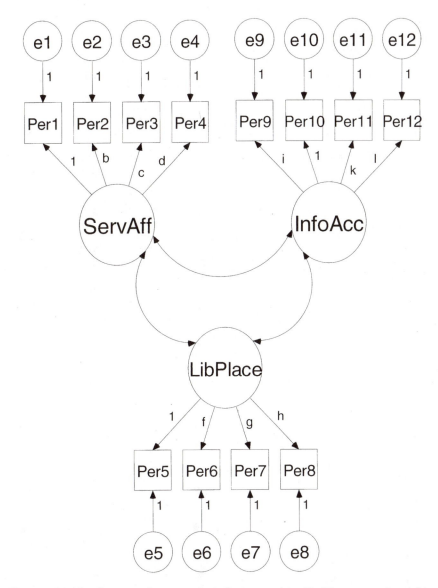

Figure 12.2. Confirmatory factor analysis factor model with 12 measured variables and 3 correlated factors with equality constraints. ServAff = "Service Affect"; InfoAcc = "Information Access"; and LibPlace = "Library as Place."

In other words, the equality constraint means that the Figure 12.2 model is more parsimonious than the Figure 12.1 model by a factor of 9 degrees of freedom (111 – 102).

Table 12.2 presents the standardized factor pattern coefficients, the factor structure coefficients, the factor correlations, and fit statistics for the

TABLE 12.2
Factor Pattern/Structure Coefficients and Fit Statistics for First-Order
Factor Analysis With Equality Constraints on Unstandardized Factor
Pattern Coefficients (n_1 = 100 Graduate Students; n_2 = 100 Faculty)

Group/ variable/ statistic	Pattern			Structure		
	I	II	III	I	II	III
Graduate students						
PER1	(0.944)	—	—	0.944	0.536	0.605
PER2	0.893	—	—	0.893	0.507	0.572
PER3	0.812	—	—	0.812	0.461	0.520
PER4	0.834	—	—	0.834	0.474	0.535
PER5	—	(0.907)	—	0.515	0.907	0.513
PER6	—	0.789	—	0.448	0.789	0.446
PER7	—	0.884	—	0.502	0.884	0.500
PER8	—	0.816	—	0.463	0.816	0.461
PER9	—	—	0.825	0.529	0.466	0.825
PER10	—	—	(0.763)	0.489	0.431	0.763
PER11	—	—	0.798	0.511	0.451	0.798
PER12	—	—	0.607	0.389	0.343	0.607
$r_{I \times II}$	0.568					
$r_{I \times III}$	0.641					
$r_{II \times III}$	0.565					
Faculty						
PER1	(0.947)	—	—	0.947	0.400	0.662
PER2	0.870	—	—	0.870	0.367	0.608
PER3	0.735	—	—	0.735	0.310	0.514
PER4	0.874	—	—	0.874	0.369	0.611
PER5	—	(0.932)	—	0.393	0.932	0.562
PER6	—	0.898	—	0.379	0.898	0.542
PER7	—	0.935	—	0.395	0.935	0.564
PER8	—	0.814	—	0.344	0.814	0.491
PER9	—	—	0.589	0.412	0.355	0.589
PER10	—	—	(0.658)	0.460	0.397	0.658
PER11	—	—	0.690	0.482	0.416	0.690
PER12	—	—	0.488	0.341	0.294	0.488
$r_{I \times II}$	0.422					
$r_{I \times III}$	0.699					
$r_{II \times III}$	0.603					
χ^2	144.47					
df	111					
χ^2/df	1.30					
NFI	0.919					
CFI	0.980					
RMSEA	0.039					

Note. Parameter estimates "fixed" to be zeroes are reported as dashes ("—"). Parameter estimates "fixed" to equal 1.0 for the unstandardized model are reported in parentheses. All critical ratios > |2.0|. NFI = normed fit index; CFI = comparative fit index; RMSEA = root-mean-square error of approximation.

analysis. The fit statistics suggest that the model is again a plausible fit to the data provided by both groups.

The fit is not as good as the fit of the less constrained model, but nevertheless remains plausible. Further restrictions on equalities might be imposed (e.g., constraining factor covariances to be equal), but such additional constraints would necessitate some reasons as to why each set of additional parameters might be expected to be invariant.

Alternatively, a sequence of equality constraints might be imposed to detect where, if at all, invariance breaks down across groups (Byrne, 2001; Jöreskog & Sörbom, 1993). The sequence of invariance tests with equality constraints might proceed as follows: (a) equality of unstandardized pattern coefficients; (b) equality of unstandardized pattern coefficients and factor covariances; (c) equality of unstandardized pattern coefficients, factor covariances, and factor variances; and (d) equality of unstandardized pattern coefficients, factor covariances, factor variances, and error variances of measured variables.

SAME MODEL, TEST INVARIANCE OF PARAMETERS

The ultimate test of model invariance is using some or all of the specific parameter estimates from one sample in a model test with an independent sample. Here the unstandardized pattern coefficients for the model estimated with graduate students were fit to the data provided by the faculty. Figure 12.3 presents the input model.

Table 12.3 presents the standardized factor pattern coefficients, the factor structure coefficients, the factor correlations, and fit statistics for the analysis. The fit statistics suggest that the model is near the boundary of plausible fit to the data provided by the faculty. Nevertheless, the degree of fit is striking given that selected *specific parameters* from the graduate students were used in the tested model.

In practice, researchers usually hope that the same constructs are measured across groups but usually do not expect factor pattern coefficients to be completely invariant. Nevertheless, such tests illustrate the power of CFA to find the boundaries of result invariance.

MAJOR CONCEPTS

Science is about isolating and understanding latent constructs. We seek constructs that replicate understated conditions and that are useful for stated purposes. In other words, we seek factors that are invariant over samples and analyses.

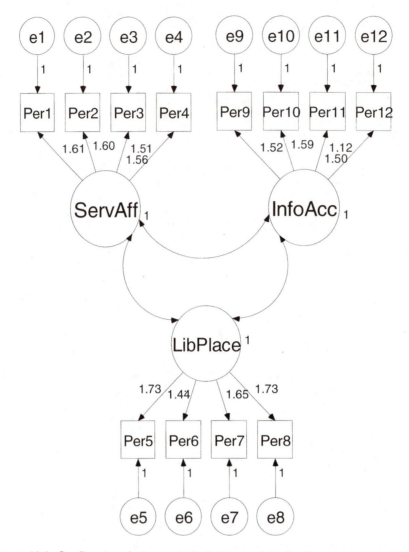

Figure 12.3. Confirmatory factor analysis factor model with 12 measured variables and 3 correlated factors with 12 unstandardized pattern coefficients from graduate students constrained as fixed. ServAff = "Service Affect"; InfoAcc = "Information Access"; and LibPlace = "Library as Place."

For example, do we obtain the same or similar factors across independent samples of people? Do we obtain similar factors when we analyze two similar but different sets of measured variables? Do we obtain similar factors when we invoke different estimation theories?

CFA is ideally suited to testing invariance propositions, because of its flexibility and capacity to quantify degrees of model fit. When rival models for a given data set are nested, we can quantify the degree to which more

TABLE 12.3
Factor Pattern/Structure Coefficients and Fit Statistics for First-Order Factor Analysis for Faculty Data ($n = 100$) Fitting Graduate Students' Unstandardized Factor Pattern Coefficients

Variable/ statistic	Pattern			Structure		
	I	II	III	I	II	III
PER1	[0.966]	—	—	0.966	0.445	0.756
PER2	[0.902]	—	—	0.902	0.415	0.706
PER3	[0.816]	—	—	0.816	0.375	0.638
PER4	[0.901]	—	—	0.901	0.415	0.705
PER5	—	[0.917]	—	0.422	0.917	0.527
PER6	—	[0.837]	—	0.385	0.837	0.481
PER7	—	[0.929]	—	0.428	0.929	0.534
PER8	—	[0.801]	—	0.369	0.801	0.461
PER9	—	—	[0.708]	0.554	0.407	0.708
PER10	—	—	[0.753]	0.590	0.433	0.753
PER11	—	—	[0.808]	0.632	0.464	0.808
PER12	—	—	[0.613]	0.480	0.352	0.613
$r_{I \times II}$	0.460					
$r_{I \times III}$	0.783					
$r_{II \times III}$	0.575					
χ^2	106.71					
df	63					
χ^2/df	1.69					
NFI	0.879					
CFI	0.946					
RMSEA	0.084					

Note. Parameter estimates "fixed" to be zeroes are reported as dashes ("—"). Parameter estimates "fixed" to equal unstandardized pattern coefficients for the graduate students are reported in brackets. All critical ratios > |2.0|. NFI = normed fit index; CFI = comparative fit index; RMSEA = root-mean-square error of approximation. 78–(12 + 13) "freed" estimates = 36.

parsimonious models estimating fewer parameters fit less well, as noted in chapter 11. Remember that less parsimonious models (i.e., models estimating more parameters) fit as well as or better than more parsimonious models, until at the extreme we estimate all possible parameters (i.e., $df = 0$), at which point every just-identified model with zero degrees of freedom fits perfectly.

Because χ^2s for nested models are additive, we can subtract the χ^2 for the less parsimonious model (e.g., $\chi^2 = 31.36$, $df = 19$) from the χ^2 for the more parsimonious model (e.g., $\chi^2 = 48.27$, $df = 20$) to quantify how much better the less parsimonious model fits (e.g., χ^2 $48.27 - 31.36 = 16.91$). Because degrees of freedom from nested models are additive, we can also subtract the degrees of freedom (e.g., $20 - 19 = 1$). This allows calculation of the p value (e.g., $p = .00004$ for $\chi^2 = 16.91$ with $df = 1$) for the null hypothesis that the nested models fit equally well.

In CFA, we can fit similar but different models to different types of samples or to the same sample over time. We can fit the same models to different samples but free the parameters to be estimated independently in each data set. Or we can fit the same model and also use some or all of the estimates in one sample with the data from an independent sample (see Thompson et al., 2003). Analyses of the last kind are the ultimate in tests of invariance, and the ultimate in parsimony, because at the extreme we are estimating no parameters with new data when we fit only the parameters from a prior study.

APPENDIX A
LibQUAL+™ Data

```
001 2 8 7 5 5 3 2 3 3 6 5 5 7
002 2 5 7 5 5 4 5 5 6 3 4 3 6
003 2 6 5 5 6 5 3 5 3 3 5 4 7
004 2 5 5 4 6 4 4 4 4 2 1 5 4
005 2 5 5 5 5 5 4 6 2 4 4 5 7
006 2 7 7 7 8 7 8 7 6 6 8 7 5
007 2 8 8 7 7 6 7 6 7 6 5 6 5
008 2 7 7 7 7 5 3 3 6 5 4 5 8
009 2 1 3 1 1 1 1 1 1 1 1 1 1
010 2 9 9 9 8 5 7 5 2 7 6 7 7
011 2 9 9 9 7 7 9 9 7 4 6 5 9
012 2 4 3 3 5 6 5 4 5 2 5 1 4
013 2 6 6 5 7 7 7 7 7 6 5 6 7
014 2 7 7 2 7 2 3 2 3 6 6 5 9
015 2 9 9 9 9 7 5 4 5 4 1 1 4
016 2 9 9 9 9 5 5 5 7 6 7 6 6
017 2 8 8 9 9 7 6 6 8 6 8 9 9
018 2 8 5 8 5 4 8 5 8 4 3 4 7
019 2 9 9 9 9 6 7 6 6 7 9 9 9
020 2 8 7 7 9 5 6 6 4 5 6 2 9
021 2 9 9 8 9 9 9 9 8 8 6 6 7
022 2 7 7 7 8 7 6 7 6 7 8 8 7
023 2 7 8 7 6 5 6 5 6 6 5 7 7
024 2 8 9 7 8 5 3 4 5 9 9 7 7
025 2 8 8 8 8 6 6 6 7 9 6 9 9
026 2 8 9 9 9 6 6 6 6 8 8 9 8
027 2 8 8 8 8 8 7 8 8 8 9 8 7
028 2 9 9 9 7 6 7 7 6 9 9 6 9
029 2 6 5 6 5 5 6 6 5 4 5 4 5
030 2 8 8 7 8 7 8 8 7 8 7 6 7
031 2 7 7 7 8 8 7 8 7 7 6 8 8
032 2 1 1 2 1 1 1 1 1 6 6 5 9
033 2 8 8 8 8 8 7 7 8 6 1 6 7
034 2 8 8 6 7 5 6 5 6 8 9 8 7
035 2 8 8 8 5 7 4 8 5 6 6 6 7
036 2 8 8 8 8 5 5 6 5 7 7 7 8
037 2 6 4 5 6 7 7 7 8 5 5 4 4
```

```
038 2 5 4 5 3 3 4 3 4 6 7 3 3
039 2 5 7 6 7 7 5 6 7 6 6 5 5
040 2 7 7 7 8 6 7 7 7 7 6 8 8
041 2 6 8 7 7 8 7 7 7 7 9 7 7
042 2 7 8 8 5 6 6 6 6 6 5 5 6
043 2 8 8 7 7 5 6 5 6 6 7 6 7
044 2 7 7 5 7 4 5 4 4 8 7 6 8
045 2 9 9 9 9 8 9 8 8 9 7 7 9
046 2 7 7 7 7 7 7 7 7 7 7 7 1
047 2 7 5 7 6 4 7 5 4 7 9 6 7
048 2 1 3 3 1 4 5 4 4 9 6 7 7
049 2 7 8 6 7 8 8 8 8 7 6 7 7
050 2 8 8 7 8 8 7 7 8 7 8 7 7
051 2 9 8 8 8 6 4 6 8 6 7 6 7
052 2 9 9 8 9 8 7 8 7 8 7 9 9
053 2 6 6 5 6 8 4 6 7 7 7 6 7
054 2 8 8 8 4 5 4 5 8 7 9 6 7
055 2 8 8 7 9 2 2 2 1 5 2 6 8
056 2 8 8 6 8 3 3 5 6 6 5 4 7
057 2 9 9 9 9 2 2 2 2 8 9 9 9
058 2 8 8 8 8 7 7 8 8 8 8 8 8
059 2 6 5 6 7 6 7 6 7 5 6 6 7
060 2 8 9 7 7 6 5 6 6 7 7 6 7
061 2 7 7 7 7 7 7 7 8 7 7 7 7
062 2 6 7 7 6 4 4 5 7 6 3 7 3
063 2 8 8 7 8 8 6 7 9 7 7 7 8
064 2 8 7 8 8 4 4 4 3 8 2 6 4
065 2 8 9 8 7 7 7 7 7 6 6 7 7
066 2 8 8 9 8 8 8 8 8 7 9 6 7
067 2 7 7 7 7 7 7 6 8 7 6 7 8
068 2 6 3 7 9 4 3 3 3 9 4 5 5
069 2 9 8 8 9 8 8 7 8 6 7 6 9
070 2 6 6 6 5 4 4 4 4 5 2 5 6
071 2 8 8 9 8 7 7 7 7 8 6 7 8
072 2 8 8 8 8 7 6 8 7 7 7 7 8
073 2 9 9 7 8 4 5 6 7 6 7 6 7
074 2 8 8 8 8 6 6 6 6 8 8 8 8
075 2 8 8 8 7 8 7 8 7 6 8 8 9
076 2 8 8 7 8 4 5 6 8 8 5 6 7
077 2 6 6 6 6 7 7 5 6 6 5 3 7
078 2 7 5 5 4 3 5 5 6 2 2 6 8
079 2 4 4 5 8 7 8 8 8 6 4 6 9
080 2 9 9 8 7 8 1 5 9 9 6 9 9
```

```
081 2 8 5 8 7 3 1 6 1 7 7 6 9
082 2 9 7 7 9 9 8 7 9 8 8 8 7
083 2 8 8 8 8 6 4 7 7 7 9 6 6
084 2 7 7 7 7 6 6 6 7 6 7 6 8
085 2 4 2 3 4 1 4 1 1 1 1 1 1
086 2 6 7 5 6 5 6 6 6 5 6 6 7
087 2 6 6 7 7 8 5 6 9 6 7 6 7
088 2 8 8 7 8 6 5 7 7 8 5 6 9
089 2 9 9 9 9 9 6 9 9 8 9 9 7
090 2 5 5 5 5 3 3 4 4 7 5 5 7
091 2 8 8 8 8 7 7 7 7 6 9 7 9
092 2 5 5 4 5 7 7 7 7 7 6 7 5
093 2 8 8 8 8 6 6 6 6 4 4 6 7
094 2 8 8 9 8 8 6 8 8 9 5 7 9
095 2 8 7 7 7 8 6 7 6 7 7 7 8
096 2 8 8 8 8 7 7 7 6 7 5 7 8
097 2 7 8 7 8 7 5 8 5 7 8 7 7
098 2 4 5 2 1 5 1 2 2 1 2 6 2
099 2 8 8 1 8 3 5 8 3 4 2 1 7
100 2 9 8 8 9 7 4 9 8 9 8 9 9
101 3 7 7 7 7 6 7 6 6 7 6 6 8
102 3 8 8 6 5 4 4 4 7 6 7 5 8
103 3 5 5 5 5 5 7 5 6 4 5 5 5
104 3 9 9 9 9 6 6 6 7 5 6 9 7
105 3 9 9 8 8 5 3 5 6 8 6 6 8
106 3 6 7 5 7 6 5 5 6 6 7 8 4
107 3 8 6 9 7 6 8 8 8 7 5 5 6 7
108 3 8 7 8 8 6 4 6 7 6 5 6 7
109 3 6 6 6 5 5 5 5 4 6 7 5 8
110 3 8 7 7 8 6 8 4 6 6 8 7 7
111 3 9 8 9 9 9 9 9 9 8 5 8 9
112 3 7 7 8 7 6 8 7 5 6 6 7 5
113 3 7 7 7 6 6 6 6 6 4 6 7 6
114 3 7 7 7 8 1 2 2 1 7 7 8 3
115 3 7 7 8 7 7 7 7 7 7 5 6 4
116 3 8 9 8 8 6 5 6 9 4 3 5 7
117 3 5 4 5 5 7 7 5 5 1 3 5 3
118 3 9 9 8 9 7 7 7 9 7 6 8 9
119 3 8 8 8 8 7 7 7 8 6 7 8 8
120 3 7 6 8 8 7 6 7 6 7 8 7 5
121 3 8 8 8 9 8 8 8 8 4 7 6 7
122 3 8 7 8 8 5 4 6 5 6 7 6 7
123 3 5 3 4 5 2 2 2 2 2 2 2 7
```

```
124 3 9 9 9 9 8 8 8 8 6 7 6 9
125 3 8 8 7 9 3 3 3 5 4 5 6 8
126 3 9 9 9 9 9 8 8 9 6 7 8 8
127 3 9 9 8 9 8 8 8 8 8 7 8 9
128 3 9 9 7 9 7 7 7 6 8 7 8 7
129 3 3 3 7 3 7 5 6 3 7 3 6 6
130 3 9 9 8 9 6 7 7 7 7 7 8 8
131 3 8 8 6 7 3 5 4 5 8 6 8 8
132 3 5 5 5 5 2 3 3 2 6 6 5 5
133 3 7 6 8 8 5 5 5 6 5 4 5 8
134 3 8 8 9 8 6 6 6 7 8 7 8 8
135 3 8 8 9 8 8 9 8 8 7 3 7 7
136 3 7 6 8 8 4 4 4 6 8 6 7 8
137 3 8 8 8 8 2 2 2 2 1 1 6 4
138 3 8 7 8 9 6 7 5 7 6 8 6 6
139 3 7 8 8 7 6 7 6 7 4 8 8 9
140 3 9 9 7 8 7 8 7 8 5 6 8 8
141 3 8 7 7 8 6 7 8 7 7 6 6 7
142 3 8 7 6 7 6 6 6 6 7 2 5 8
143 3 7 7 7 5 3 2 5 3 9 3 7 4
144 3 8 8 8 8 9 8 8 8 8 8 8 7
145 3 8 7 8 7 3 5 5 5 6 8 5 6
146 3 7 7 6 7 6 5 6 7 7 5 7 6
147 3 8 8 7 8 5 6 5 6 6 3 5 3
148 3 5 7 5 7 6 6 6 6 5 5 6 7
149 3 9 9 9 9 7 7 7 7 9 9 9 7
150 3 8 8 7 8 5 5 5 7 1 5 6 8
151 3 8 5 8 8 8 8 8 8 7 2 6 6
152 3 9 9 9 9 5 6 5 9 9 8 8 5
153 3 9 9 8 9 8 8 8 9 8 9 8 8
154 3 8 8 8 8 7 7 7 8 8 8 7 7
155 3 6 5 5 5 3 2 3 2 1 2 5 3
156 3 5 5 3 5 5 5 5 7 5 3 7 7
157 3 7 6 7 7 3 3 4 4 3 6 7 4
158 3 7 6 6 7 5 5 7 6 7 4 8 6
159 3 9 8 9 9 6 6 6 7 7 9 8 9
160 3 7 7 7 7 6 6 6 2 8 8 6 7
161 3 8 8 3 7 5 5 5 4 7 5 6 7
162 3 8 8 8 8 8 8 8 9 7 3 7 2
163 3 9 9 9 9 1 1 1 1 6 5 6 9
164 3 9 9 9 9 6 6 7 6 6 3 8 7
165 3 7 7 6 7 6 7 6 6 6 7 5 6
166 3 9 5 9 8 9 9 9 3 9 8 9 7
```

```
167 3 9 7 9 7 8 7 8 7 5 7 6 7
168 3 5 3 6 3 3 2 3 5 6 2 2 2
169 3 9 8 7 7 7 7 7 9 7 8 7 6
170 3 9 9 7 9 1 1 3 2 4 3 8 5
171 3 8 8 7 7 5 7 7 7 7 4 4 7
172 3 6 7 5 6 6 6 6 6 4 5 6 4
173 3 9 8 9 8 9 9 9 9 6 7 6 7
174 3 9 9 8 9 2 5 5 6 7 5 8 7
175 3 6 6 7 8 6 5 6 5 6 7 7 6
176 3 8 8 5 9 7 3 3 3 5 7 8 7
177 3 9 9 8 9 9 8 9 9 7 6 6 5
178 3 8 8 7 8 3 3 7 4 5 5 5 7
179 3 7 7 7 8 5 5 5 6 6 6 7 7
180 3 8 8 8 7 4 4 4 7 5 8 8 7
181 3 8 8 8 9 5 8 5 6 8 8 6 7
182 3 8 8 8 8 6 6 6 6 7 8 8 7
183 3 9 9 9 9 8 8 8 8 9 7 7 8
184 3 9 7 5 7 5 6 5 5 4 5 3 8
185 3 9 9 9 9 8 7 8 7 5 7 9 9
186 3 6 7 5 4 5 5 5 3 5 5 5 5
187 3 9 8 8 9 7 6 7 7 6 7 7 7
188 3 9 9 9 9 8 8 8 8 6 6 7 8
189 3 6 5 6 7 1 2 1 2 2 4 6 5
190 3 9 9 9 9 8 7 7 9 8 9 9 9
191 3 9 9 9 9 4 5 4 4 4 5 6 9
192 3 9 9 8 7 3 3 3 5 5 7 7 9
193 3 9 9 6 9 8 5 5 9 4 9 8 7
194 3 8 7 6 6 7 8 7 9 6 8 9 8
195 3 6 6 5 6 7 6 7 8 6 7 9 6
196 3 6 6 4 5 5 4 5 5 6 4 2 6
197 3 9 9 9 9 9 6 7 9 6 6 6 7
198 3 6 5 5 5 3 5 3 3 3 3 6 7
199 3 6 4 4 6 8 8 8 8 8 7 8 6
200 3 4 4 7 4 2 1 2 2 2 3 6 8
```

Note. The first column reports case ID numbers. The second column represents ROLETYPE ("2" = graduate students; "3" = faculty). The next 12 variables (PER1 to PER12) are the 12 rating criteria presented in Table 3.1.

```
SET PRINTBACK=LISTING .
COMMENT Type in the pattern matrix that is the target as
  matrix "A"; Then type in the pattern matrix to be
  rotated to 'best-fit' position with the target as
  matrix "B"; Type in a matrix "N_A" consisting of all
  zeroes with the same number or rows as "A" and "B";
  Type in a square matrix of all zeroes with the number
  of rows and columns equal to the number of orthogonal
  factors in "A" and "B" .
MATRIX .
COMMENT 100 Grad Students, Varimax Pattern Matrix .
COMPUTE A =
{ .92134, .18690, .18737 ;
 .84248, .21945, .25625 ;
 .75606, .28211, .35335 ;
 .80837, .28360, .19958 ;
 .22292, .84662, .26935 ;
 .17110, .85348, .03794 ;
 .29969, .82875, .22423 ;
 .22454, .80378, .26075 ;
 .30781, .13580, .81976 ;
 .15795, .23077, .80478 ;
 .27222, .20840, .80207 } .
COMMENT 100 Faculty, Varimax Pattern Matrix .
COMPUTE B =
{ .20284, .89659, .18675 ;
 .10828, .87403, .21537 ;
 .23051, .73176, .16198 ;
 .16361, .87285, .24099 ;
 .90807, .12646, .21722 ;
 .89654, .13470, .21639 ;
 .90832, .17788, .19870 ;
 .79337, .32388, .14477 ;
 .33664, .10530, .66769 ;
 .18185, .22943, .79192 ;
 .11897, .33942, .71967 } .
COMPUTE N_A =
{ .0 ;
 .0 ;
 .0 ;
 .0 ;
 .0 ;
```

```
        .0 ;
        .0 ;
        .0 ;
        .0 ;
        .0 ;
        .0 } .
COMPUTE DIAG_M =
{ .0, .0, .0 ;
  .0, .0, .0 ;
  .0, .0, .0 } .
COMPUTE N_B=N_A .
PRINT A /
 FORMAT='F8.2' /
 TITLE='First Pattern Matrix (Target)' /
 SPACE=4 /
 RLABELS=Per1, Per2, Per3, Per4, Per5, Per6,
 Per7, Per8, Per9, Per10, Per11 /
 CLABELS=Fact_I, Fact_II, Fact_III / .
COMPUTE A_N=A .
-LOOP #I=1 TO NROW(A) .
+ LOOP #J=1 TO NCOL(A) .
COMPUTE A_N(#I,#J)=A(#I,#J) ** 2 .
+ END LOOP .
-END LOOP .
PRINT A_N /
 FORMAT='F8.4' /
 TITLE='First Pattern Matrix (Target) Squared' /
 SPACE=4 /
 RLABELS=Per1, Per2, Per3, Per4, Per5, Per6,
 Per7, Per8, Per9, Per10, Per11 /
 CLABELS=Fact_I, Fact_II, Fact_III / .
-LOOP #J=1 TO NCOL(A) .
+ LOOP #I=1 TO NROW(A) .
COMPUTE N_A(#I)=A_N(#I,#J) + N_A(#I) .
+ END LOOP .
-END LOOP .
PRINT N_A /
 FORMAT='F8.3' /
 TITLE='Row Sum of Squares for First Pattern Matrix' /
 SPACE=4 /
 RLABELS=Per1, Per2, Per3, Per4, Per5, Per6,
 Per7, Per8, Per9, Per10, Per11 / .
LOOP #I=1 TO NROW(A) .
COMPUTE N_A(#I) = 1.0 / (N_A(#I) ** .5) .
END LOOP .
PRINT N_A /
FORMAT='F8.3' /
TITLE='Normalization Factor for Rows' /
SPACE=4 /
RLABELS=Per1, Per2, Per3, Per4, Per5, Per6,
```

```
Per7, Per8, Per9, Per10, Per11 / .
-LOOP #J=1 TO NCOL(A) .
+ LOOP #I=1 TO NROW(A) .
COMPUTE A_N(#I,#J)=A(#I,#J) * N_A(#I) .
+ END LOOP .
-END LOOP .
PRINT A_N /
FORMAT='F8.4' /
TITLE='First Pattern Matrix (Target) Normalized' /
SPACE=4 /
RLABELS=Per1, Per2, Per3, Per4, Per5, Per6,
Per7, Per8, Per9, Per10, Per11 /
CLABELS=Fact_I, Fact_II, Fact_III / .
PRINT B /
FORMAT='F8.2' /
TITLE='Second Pattern Matrix' /
SPACE=4 /
RLABELS=Per1, Per2, Per3, Per4, Per5, Per6,
Per7, Per8, Per9, Per10, Per11 /
CLABELS=Fact_I, Fact_II, Fact_III / .
COMPUTE B_N=B .
-LOOP #I=1 TO NROW(B) .
+ LOOP #J=1 TO NCOL(B) .
COMPUTE B_N(#I,#J)=B(#I,#J) ** 2 .
+ END LOOP .
-END LOOP .
PRINT B_N /
FORMAT='F8.4' /
TITLE='Second Pattern Matrix Squared' /
SPACE=4 /
RLABELS=Per1, Per2, Per3, Per4, Per5, Per6,
Per7, Per8, Per9, Per10, Per11 /
CLABELS=Fact_I, Fact_II, Fact_III / .
-LOOP #J=1 TO NCOL(B) .
+ LOOP #I=1 TO NROW(B) .
COMPUTE N_B(#I)=B_N(#I,#J) + N_B(#I) .
+ END LOOP .
-END LOOP .
PRINT N_B /
FORMAT='F8.3' /
TITLE='Row Sum of Squares for Second Pattern Matrix' /
SPACE=4 /
RLABELS=Per1, Per2, Per3, Per4, Per5, Per6,
Per7, Per8, Per9, Per10, Per11 / .
LOOP #I=1 TO NROW(B) .
COMPUTE N_B(#I) = 1.0 / (N_B(#I) ** .5) .
END LOOP .
PRINT N_B /
FORMAT='F8.3' /
TITLE='Normalization Factor for Rows' /
```

```
SPACE=4 /
RLABELS=Per1, Per2, Per3, Per4, Per5, Per6,
Per7, Per8, Per9, Per10, Per11 / .
-LOOP #J=1 TO NCOL(B) .
+ LOOP #I=1 TO NROW(B) .
COMPUTE B_N(#I,#J)=B(#I,#J) * N_B(#I) .
+ END LOOP .
-END LOOP .
PRINT B_N /
FORMAT='F8.4' /
TITLE='Second Pattern Matrix Normalized' /
SPACE=4 /
RLABELS=Per1, Per2, Per3, Per4, Per5, Per6,
Per7, Per8, Per9, Per10, Per11 /
CLABELS=Fact_I, Fact_II, Fact_III / .
COMPUTE A_T=TRANSPOS(A_N) .
PRINT A_T /
FORMAT='F8.2' /
TITLE='A_N Transpose' /
SPACE=4 /
RLABELS=Fact_I, Fact_II, Fact_III /
CLABELS=Per1, Per2, Per3, Per4, Per5, Per6,
Per7, Per8, Per9, Per10, Per11 / .
COMPUTE B_T=TRANSPOS(B_N) .
PRINT B_T /
FORMAT='F8.2' /
TITLE='B_N Transpose' /
SPACE=4 /
RLABELS=Fact_I, Fact_II, Fact_III /
CLABELS=Per1, Per2, Per3, Per4, Per5, Per6,
Per7, Per8, Per9, Per10, Per11 / .
COMPUTE RI=A_T * B_N .
PRINT RI /
FORMAT='F8.3' /
TITLE='A_N Transpose times B_N' /
SPACE=4 /
RLABELS=Fact_I, Fact_II, Fact_III /
CLABELS=Fact_I, Fact_II, Fact_III / .
COMPUTE RI_T=TRANSPOS(RI) .
PRINT RI_T /
FORMAT='F8.3' /
TITLE='Transpose of (A_N Transpose times B_N)' /
SPACE=4 /
RLABELS=Fact_I, Fact_II, Fact_III /
CLABELS=Fact_I, Fact_II, Fact_III / .
COMPUTE QUAD=RI * RI_T .
PRINT QUAD /
FORMAT='F8.3' /
TITLE='A_N Trans * B_N * Trans of (A_N Trans * B_N)' /
SPACE=4 /
```

```
RLABELS=Fact_I, Fact_II, Fact_III / 
CLABELS=Fact_I, Fact_II, Fact_III / .
CALL EIGEN(QUAD, EIGVEC, EIG) .
PRINT EIG /
FORMAT='F8.3' /
TITLE='Eigenvalues of QUAD' /
SPACE=4 /
RLABELS=Fact_I, Fact_II, Fact_III / .
PRINT EIGVEC /
FORMAT='F8.3' /
TITLE='Eigenvectors of QUAD' /
SPACE=4 /
RLABELS=ONE, TWO /
CLABELS=Fact_I, Fact_II, Fact_III / .
-LOOP #I=1 TO NROW(QUAD) .
+ LOOP #J=1 TO NROW(QUAD) .
COMPUTE EIGVEC(#I,#J)=EIGVEC(#I,#J) * (EIG(#J) ** .5) .
+ END LOOP .
-END LOOP .
PRINT EIGVEC /
FORMAT='F8.3' /
TITLE='Pattern Coefficients of QUAD' /
SPACE=4 /
RLABELS=ONE, TWO /
CLABELS=Fact_I, Fact_II, Fact_III / .
LOOP I=1 TO NROW(EIG) .
COMPUTE EIG(I)=EIG(I) ** -1.5 .
END LOOP .
PRINT EIG /
FORMAT='F8.3' /
TITLE='Eigenvalues raised to -1.5' /
SPACE=4 /
RLABELS=Fact_I, Fact_II, Fact_III / .
CALL SETDIAG(DIAG_M,EIG) .
PRINT DIAG_M /
FORMAT='F8.3' /
TITLE='Diagonal Matrix (Eigenvalues raised to -1.5)' /
SPACE=4 /
CLABELS=Fact_I, Fact_II, Fact_III /
RLABELS=Fact_I, Fact_II, Fact_III / .
COMPUTE VEC_T=TRANSPOS(EIGVEC) .
PRINT VEC_T /
FORMAT='F8.3' /
TITLE='Transpose of Eigenvectors' /
SPACE=4 /
RLABELS=Fact_I, Fact_II, Fact_III /
CLABELS=ONE, TWO / .
COMPUTE D=RI_T * EIGVEC .
PRINT D /
FORMAT='F9.3' /
```

```
TITLE='D= trans (trans A times B) times Eigenvectors' /
SPACE=4 /
RLABELS=Fact_I, Fact_II, Fact_III /
CLABELS=Fact_I, Fact_II, Fact_III / .
-LOOP J=1 TO NCOL(A) .
COMPUTE EE=EIG(J) .
+ LOOP I=1 TO NCOL(A) .
COMPUTE D(I,J)=D(I,J) * EE .
+ END LOOP .
-END LOOP .
PRINT D /
FORMAT='F9.3' /
TITLE='D = D times Eigenvalues ** -1.5' /
SPACE=4 /
RLABELS=Fact_I, Fact_II, Fact_III /
CLABELS=Fact_I, Fact_II, Fact_III / .
COMPUTE D_T=TRANSPOS(D) .
PRINT D_T /
FORMAT='F9.3' /
TITLE='D transposed' /
SPACE=4 /
RLABELS=Fact_I, Fact_II, Fact_III /
CLABELS=Fact_I, Fact_II, Fact_III / .
COMPUTE C=EIGVEC * D_T .
PRINT C /
FORMAT='F9.3' /
TITLE='Factor Correlations (Cosines)' /
SPACE=4 /
RLABELS=Fact_Ia, Fact_IIa, Fact_IIIa /
CLABELS=Fact_Ib, Fact_IIb, Fact_IIIb / .
COMPUTE C=D * VEC_T .
COMPUTE B_ROT=B * C .
PRINT B_ROT /
FORMAT='F8.3' /
TITLE='B rotated to Best-Fit with A' /
SPACE=4 /
RLABELS=Per1, Per2, Per3, Per4, Per5, Per6,
Per7, Per8, Per9, Per10, Per11 /
CLABELS=Fact_I, Fact_II, Fact_III / .
COMPUTE BROT_N=B_ROT .
-LOOP #I=1 TO NROW(A) .
+ LOOP #J=1 TO NCOL(A) .
COMPUTE BROT_N(#I,#J)=B_ROT(#I,#J) ** 2 .
+ END LOOP .
COMPUTE N_A(#I)= .0 .
-END LOOP .
PRINT BROT_N /
FORMAT='F8.4' /
TITLE='Best Fit Pattern Matrix (Target) Squared' /
SPACE=4 /
```

```
RLABELS=Per1, Per2, Per3, Per4, Per5, Per6,
Per7, Per8, Per9, Per10, Per11 /
CLABELS=Fact_I, Fact_II, Fact_III / .
-LOOP #J=1 TO NCOL(A) .
+ LOOP #I=1 TO NROW(A) .
COMPUTE N_A(#I)=BROT_N(#I,#J) + N_A(#I) .
+ END LOOP .
-END LOOP .
PRINT N_A /
FORMAT='F8.3' /
TITLE='Row Sum of Squares for Best Fit Matrix' /
SPACE=4 /
RLABELS=Per1, Per2, Per3, Per4, Per5, Per6,
Per7, Per8, Per9, Per10, Per11 / .
LOOP #I=1 TO NROW(A) .
COMPUTE N_A(#I) = 1.0 / (N_A(#I) ** .5) .
END LOOP .
PRINT N_A /
FORMAT='F8.3' /
TITLE='Normalization Factor for Rows' /
SPACE=4 /
RLABELS=Per1, Per2, Per3, Per4, Per5, Per6,
Per7, Per8, Per9, Per10, Per11 / .
-LOOP #J=1 TO NCOL(A) .
+ LOOP #I=1 TO NROW(A) .
COMPUTE BROT_N(#I,#J)=B_ROT(#I,#J) * N_A(#I) .
+ END LOOP .
-END LOOP .
PRINT BROT_N /
FORMAT='F8.4' /
TITLE='Best Fit Pattern Matrix (Target) Normalized' /
SPACE=4 /
RLABELS=Per1, Per2, Per3, Per4, Per5, Per6,
Per7, Per8, Per9, Per10, Per11 /
CLABELS=Fact_I, Fact_II, Fact_III / .
COMPUTE BROTN_T=TRANSPOS(BROT_N) .
COMPUTE T_M=A_N * BROTN_T .
COMPUTE TEST=DIAG(T_M) .
PRINT TEST /
FORMAT='F8.3' /
TITLE='Test Vector Cosines for Variables' /
SPACE=4 /
RLABELS=Per1, Per2, Per3, Per4, Per5, Per6,
Per7, Per8, Per9, Per10, Per11 / .
END MATRIX .
```

NOTATION

$\mathbf{C}_{V \times V}$ The symmetric variance/covariance (or simply covariance) matrix with variances on the diagonal and covariances of measured variables with each other off the diagonal.

D^2_i The Mahalanobis distance of the ith person's set of scores from the set of means on the measured variables.

$\mathbf{F}_{N \times V}$ The scores on the latent factors.

$\mathbf{I}_{F \times F}$ The identity matrix which, when postmultiplied times an original conformable matrix, yields as an answer the original matrix (like the number 1 in algebra).

$\mathbf{P}_{V \times F}$ The first-order factor pattern coefficient matrix.

$\mathbf{P}_{F \times H}$ The second-order factor pattern coefficient matrix.

$\mathbf{P}_{V \times H}$ The product (first-order by second-order) factor pattern coefficient matrix.

$\mathbf{R}_{F \times F}$ The symmetric matrix of bivariate correlations (e.g., Pearson r's) among the $\underline{f}$ factors.

$\mathbf{R}_{V \times V}$ The symmetric matrix of bivariate correlations (e.g., Pearson r's) among the v measured variables.

$\mathbf{R}_{V \times V}{}^{+}$ The "reproduced" correlation matrix (i.e., the portion of the intervariable correlation matrix that the factors [i.e., the pattern coefficients] *can* reproduce).

$\mathbf{R}_{V \times V}{}^{-}$ The "residual" correlation matrix (i.e., the portion of the intervariable correlation matrix that the factors [i.e., the pattern coefficients] *cannot* reproduce).

$\mathbf{R}_{V \times V}{}^{-1}$ The inverted intervariable correlation matrix, which satisfies the equation: $\mathbf{R}_{V \times V} \, \mathbf{R}_{V \times V}{}^{-1} = \mathbf{I}_{V \times V}$.

$\mathbf{S}_{V \times F}$ The first-order factor structure coefficient matrix.

$\mathbf{W}_{V \times F}$ The factor score coefficient matrix.

$\mathbf{Z}_{N \times V}$ The measured variables in z score form, such that the columns have means of zero and standard deviations (and variances) of one.

GLOSSARY

Communality coefficient (h^2). A statistic in a squared metric indicating how much of the variance in a measured variable the factors as a set can reproduce, or conversely, how much of the variance of a given measured variable was useful in delineating the factors as a set.

Component. A set of pattern coefficients for the measured variables that can be used to estimate factor scores, derived without iterative estimation of communality coefficients. (Some researchers consider *factor* a synonymous term, but others feel that *factor* is a term that should be reserved for weights derived when iteration is invoked.)

Critical ratio (or *t* or Wald statistic). A parameter estimate divided by its standard error, which is expected to be greater than roughly |2.0| for parameters that one expects to be nonzero.

Eigenvalues (λ). Area-world, variance-accounted-for statistics that characterize the amount of information present in a given factor or function. (Eigenvalues are also sometimes synonymously called "characteristic roots.")

Factor rotation. Graphic visual or mathematical movement of the axes measuring the factor space used in exploratory factor analysis so that the factors can be more readily interpreted.

Factor scores. The latent (synthetic or composite) variables computed on each factor (like the $\hat{Y}$ scores in multiple regression, or the discriminant function scores in descriptive discriminant analysis) that may then be used in further statistical analyses (e.g., *t* tests, analyses of variance) in place of the measured variables.

First-order factors. The pattern or pattern/structure coefficients for the factored entities (measured variables in the commonly used R-technique analysis).

General linear model (GLM). The notion that *all* statistical analyses are correlational and can yield effect sizes analogous to r^2, and apply weights to measured variables to obtain scores on the composite variables that are actually the focus of all these analyses.

Identification. A model is said to be "identified" when, for a given research problem and data set, sufficient constraints are imposed such that there is a single set of parameter estimates yielded by the analysis.

Inadmissible. A statistical estimate that is mathematically implausible (e.g., a negative variance, or an *r* that is outside the range of −1 to +1) or a solution that contains one or more estimates that are implausible.

Invariance. The degree to which a factor or a solution can be reproduced across samples, across different analytic methods, or both.

Inversion. The matrix algebra process of solving to find a matrix that when post-multiplied by the original matrix yields the identity matrix; inverted matrices are used in matrix algebra to accomplish division.

179

Iteration. A statistical process of estimating a set of results, and then successively tweaking them until two consecutive sets of estimates are sufficiently close that the iteration is deemed to have converged.

Loadings. An ambiguous and confusing slang term used by some authors for pattern coefficients, and by some authors for structure coefficients, and by some authors for both even within the same article even when the coefficients are not equal.

Measurement error. The variance in scores that is deemed unreliable; this area-world proportion can be estimated as 1.0 minus a reliability coefficient.

Modification index. The estimated decrease (improvement) in the model fit chi-square resulting from freeing a given previously fixed parameter instead to be estimated.

Oblique. An angle of more or less than 90° between some rotated factors, which means that the factors (or components) and factor scores are correlated.

Orthogonal. The 90° angle of all unrotated and some rotated factors, which statistically means the factors (or components) are perfectly uncorrelated.

Outliers. Persons with aberrant or anomalous scores on one or more variables with regard to one or more statistics.

Pattern coefficients. The weights applied to the measured variables to obtain scores on the factor analysis latent variables (called *factor scores*). These weights are analogous to the β weights in multiple regression, the standardized discriminant function coefficients in descriptive discriminant analysis, and the standardized canonical function coefficients in canonical correlation analysis.

Pattern/structure coefficients. The term for the entries in the single exploratory factor analysis coefficient matrix produced when factors are first extracted or after rotation when orthogonal rotation has been invoked, and in CFA when factors are constrained to be uncorrelated.

Procrustean rotation. A method that rotates a pattern (or pattern/structure) coefficient matrix to "best-fit" position with a theoretical or actual pattern coefficient matrix.

Reflection. The process of changing all the signs of pattern and structure coefficients on a given factor so that all or most of the largest values on the factor are positive in sign.

Sampling error. The unique, idiosyncratic variance in a given sample that does not exist (a) in the population, which has no idiosyncratic variance due to a sampling process, or (b) in other samples, which do have sampling error, but sampling error variances that are at least somewhat different.

Second-order factors. The pattern or pattern/structure coefficients for the second-order factors defined by the first-order factors.

Simple structure. A theoretical pattern of the dispersion of near-zero and large nonzero pattern coefficients that optimizes interpretability of the factors by there being (a) enough variables reflecting a construct and (b) most variables reflecting only a single factor such that the factors can be both labeled and distinguished from each other.

Structure coefficient. The Pearson product–moment correlation of a factored entity (measured variables in the commonly used R-technique analysis) with a latent factor.

REFERENCES

Anderson, T. W., & Rubin, H. (1956). Statistical inference in factor analysis. *Proceedings of the Third Berkeley Symposium on Mathematical Statistics and Probability, 5*, 111–150.

Arbuckle, J. L., & Wothke, W. (1999). *AMOS 4.0 user's guide*. Chicago: Smallwaters.

Armstrong, J. S. (1967). Derivation of theory by means of factor analysis or Tom Swift and his electric factor analysis machine. *American Statistician, 21*(5), 17–21.

Bagozzi, R. P., Fornell, C., & Larcker, D. F. (1981). Canonical correlation analysis as a special case of a structural relations model. *Multivariate Behavioral Research, 16*, 437–454.

Bandalos, D., & Finney, J. S. (2001). Item parceling issues in structural equation modeling. In G. A. Marcoulides & R. Schumacker (Eds.), *New developments and techniques in structural equation modeling* (pp. 269–295). Mahwah, NJ: Erlbaum.

Bartlett, M. S. (1937). The statistical conception of mental factors. *British Journal of Psychology, 28*, 97–104.

Bartlett, M. S. (1950). Tests of significance in factor analysis. *British Journal of Psychology, 3*(2), 77–85.

Bentler, P. M. (1990). Comparative fit indexes in structural models. *Psychological Bulletin, 107*, 238–246.

Bentler, P. M. (1994). On the quality of test statistics in covariance structure analysis: Caveat emptor. In C. R. Reynolds (Ed.), *Cognitive assessment: A multidisciplinary perspective* (pp. 237–260). New York: Plenum Press.

Bentler, P. M. (1995). *EQS structural equations program manual*. Encino, CA: Multivariate Software.

Bentler, P. M., & Bonnett, D. G. (1980). Significance tests and goodness of fit in the analysis of covariance structures. *Psychological Bulletin, 88*, 588–606.

Bentler, P. M., & Yuan, K. -H. (2000). On adding a mean structure to a covariance structure model. *Educational and Psychological Measurement, 60*, 326–339.

Bollen, K. A. (1987). Outliers and improper solutions: A confirmatory factor analysis example. *Sociological Methods and Research, 15*, 375–384.

Bollen, K. A. (1989). *Structural equations with latent variables*. New York: Wiley.

Borrello, G. M., & Thompson, B. (1990). An hierarchical analysis of the Hendrick-Hendrick measure of Lee's typology of love. *Journal of Social Behavior and Personality, 3*, 327–342.

Byrne, B. M. (1994). *Structural equation modeling with EQS and EQS/Windows: Basic concepts, applications, and programming*. Thousand Oaks, CA: Sage.

Byrne, B. M. (1998). *Structural equation modeling with LISREL, PRELIS, and SIMPLIS: Basic concepts, applications, and programming*. Mahwah, NJ: Erlbaum.

Byrne, B. M. (2001). *Structural equation modeling with AMOS: Basic concepts, applications, and programming.* Mahwah, NJ: Erlbaum.

Carr, S. (1992). A primer on the use of Q-technique factor analysis. *Measurement and Evaluation in Counseling and Development, 25,* 133–138.

Cattell, R. B. (1953). A quantitative analysis of the changes in the culture pattern of Great Britain, 1837–1937, by P-technique. *Acta Psychologia, 9,* 99–121.

Cattell, R. B. (1966a). The data box: Its ordering of total resources in terms of possible relational systems. In R. B. Cattell (Ed.), *Handbook of multivariate experimental psychology* (pp. 67–128). Chicago: Rand McNally.

Cattell, R. B. (1966b). The scree test for the number of factors. *Multivariate Behavioral Research, 1,* 245–276.

Cattell, R. B. (1978). *The scientific use of factor analysis in behavioral and life sciences.* New York: Plenum.

Cliff, N. (1987). *Analyzing multivariate data.* San Diego, CA: Harcourt Brace Jovanovich.

Cohen, J. (1968). Multiple regression as a general data-analytic system. *Psychological Bulletin, 70,* 426–443.

Cohen, J., & Cohen, P. (1983). *Applied multiple regression/correlation analysis for the behavioral sciences.* Hillsdale, NJ: Erlbaum.

Cohen, J., Cohen, P., West, S. G., & Aiken, L. S. (2003). *Applied multiple regression/correlation analysis for the behavioral sciences* (3rd ed.). Mahwah, NJ: Erlbaum.

Cook, C., Heath, F., & Thompson, B. (2001). Users' hierarchical perspectives on library service quality: A "LibQUAL+™" study. *College and Research Libraries, 62,* 147–153.

Cook, C., & Thompson, B. (2000). Higher-order factor analytic perspectives on users' perceptions of library service quality. *Library Information Science Research, 22,* 393–404.

Cook, C., & Thompson, B. (2001). Psychometric properties of scores from the Web-based LibQUAL+™ study of perceptions of library service quality. *Library Trends, 49,* 585–604.

Cooley, W. W., & Lohnes, P. R. (1971). *Multivariate data analysis.* New York: Wiley.

Courville, T., & Thompson, B. (2001). Use of structure coefficients in published multiple regression articles: β is not enough. *Educational and Psychological Measurement, 61,* 229–248.

Cronkhite, G., & Liska, J. R. (1980). The judgment of communicant acceptability. In M. R. Roloff & G. R. Miller (Eds.), *Persuasion: New directions in theory and research* (pp. 101–139). Beverly Hills, CA: Sage.

Crossman, L. L. (1996). Cross-validation analysis for the canonical case. In B. Thompson (Ed.), *Advances in social science methodology* (Vol. 4, pp. 95–106). Greenwich, CT: JAI Press.

Cudeck, R. (1989). The analysis of correlation matrices using covariance structure models. *Psychological Bulletin, 105,* 317–327.

Cumming, G., & Finch, S. (2001). A primer on the understanding, use and calculation of confidence intervals that are based on central and noncentral distributions. *Educational and Psychological Measurement, 61*, 532–575.

Davis, W. R. (1993). The FC1 rule of identification for confirmatory factor analysis: A general sufficient condition. *Sociological Methods and Research, 21*, 403–437.

Diaconis, P., & Efron, B. (1983). Computer-intensive methods in statistics. *Scientific American, 248*(5), 116–130.

Duncan, O. D. (1975). *Introduction to structural equation models.* New York: Academic Press.

Dunlap, W. P., & Landis, R. S. (1998). Interpretations of multiple regression borrowed from factor analysis and canonical correlation. *Journal of General Psychology, 125*, 397–407.

Fabrigar, L. R., Wegener, D. T., MacCallum, R. C., & Strahan, E. J. (1999). Evaluating the use of exploratory factor analysis in psychological research. *Psychological Methods, 4*, 272–299.

Fan, X., Thompson, B., & Wang, L. (1999). The effects of sample size, estimation methods, and model specification on SEM fit indices. *Structural Equation Modeling, 6*, 56–83.

Fan, X., Wang, L., & Thompson, B. (1997, March). *Effects of data nonnormality on fit indices and parameter estimates for true and misspecified SEM models.* Paper presented at the annual meeting of the American Educational Research Association, Chicago. (ERIC Document Reproduction Service No. ED 408 299)

Fouladi, R. T. (2000). Performance of modified test statistics in covariance and correlation structure analysis under conditions of multivariate nonnormality. *Structural Equation Modeling, 7*, 356–410.

Frankiewicz, R. G., & Thompson, B. (1979, April). *A comparison of strategies for measuring teacher brinkmanship behavior.* Paper presented at the annual meeting of the American Educational Research Association, San Francisco. (ERIC Document Reproduction Service No. ED 171 753)

Gorsuch, R. L. (1983). *Factor analysis* (2nd ed.). Hillsdale, NJ: Erlbaum.

Gorsuch, R. L. (2003). Factor analysis. In J. A. Schinka & W. F. Velicer (Vol. Eds.), *Handbook of psychology: Vol. 2. Research methods in psychology* (pp. 143–164). Hoboken, NJ: Wiley.

Graham, J. M., Guthrie, A. C., & Thompson, B. (2003). Consequences of not interpreting structure coefficients in published CFA research: A reminder. *Structural Equation Modeling, 10*, 142–153.

Guilford, J. P. (1946). New standards for test evaluation. *Educational and Psychological Measurement, 6*, 427–439.

Guilford, J. P. (1967). *The nature of human intelligence.* New York: McGraw-Hill.

Guadagnoli, E., & Velicer, W. (1988). Relation of sample size to the stability of component patterns. *Psychological Bulletin, 103*, 265–275.

Guttman, L. (1954). Some necessary conditions for common-factor analysis. *Psychometrika, 19*, 149–161.

Hayduk, L. A. (1987). *Structural equation modeling with LISREL: Essentials and advances*. Baltimore: Johns Hopkins University Press.

Hendrickson, A. E., & White, P. O. (1964). Promax: A quick method for rotation to oblique simple structure. *British Journal of Statistical Psychology, 17*, 65–70.

Henson, R. K. (1999). Multivariate normality: What is it and how is it assessed? In B. Thompson (Ed.), *Advances in social science methodology* (Vol. 5, pp. 193–212). Stamford, CT: JAI Press.

Heywood, H. B. (1931). On finite sequences of real numbers. *Proceedings of the Royal Statistical Society of London, 134*, 486–501.

Holzinger, K. J., & Swineford, F. (1939). *A study in factor analysis: The stability of a bi-factor solution* (No. 48). Chicago: University of Chicago Press.

Horn, J. L. (1965). A rationale and test for the number of factors in factor analysis. *Psychometrika, 30*, 179–185.

Horst, P. (1966). *Psychological measurement and prediction*. Belmont, CA: Wadsworth.

Hu, L., & Bentler, P. M. (1999). Cutoff criteria for fit indexes in covariance structure analysis: Conventional criteria versus new alternatives. *Structural Equation Modeling, 6*, 1–55.

Huberty, C. (1994). *Applied discriminant analysis*. New York: Wiley.

Huberty, C. J., & Morris, J. D. (1988). A single contrast test procedure. *Educational and Psychological Measurement, 48*, 567–578.

Jackson, D. (2001). Sample size and number of parameter estimates in maximum likelihood confirmatory factor analysis: A Monte Carlo investigation. *Structural Equation Modeling, 8*, 205–223.

Jones, H. L., Thompson, B., & Miller, A. H. (1980). How teachers perceive similarities and differences among various teaching models. *Journal of Research in Science Teaching, 17*, 321–326.

Jöreskog, K. G. (1969). A general approach to confirmatory maximum likelihood factor analysis. *Psychometrika, 34*, 183–202.

Jöreskog, K. G., & Sörbom, D. (1984). *LISREL VI user's guide* (3rd ed.). Mooresville, IN: Scientific Software.

Jöreskog, K. G., & Sörbom, D. (1993). *LISREL 8: User's reference guide*. Chicago: Scientific Software International.

Kaiser, H. F. (1958). The varimax criterion for analytic rotation in factor analysis. *Psychometrika, 23*, 187–200.

Kerlinger, F. N. (1979). *Behavioral research: A conceptual approach*. New York: Holt, Rinehart & Winston.

Kerlinger, F. N. (1984). *Liberalism and conservatism: The nature and structure of social attitudes*. Hillsdale, NJ: Erlbaum.

Kerlinger, F. N. (1986). *Foundations of behavioral research* (3rd ed.). New York: Holt, Rinehart & Winston.

Knapp, T. R. (1978). Canonical correlation analysis: A general parametric significance testing system. *Psychological Bulletin, 85,* 410–416.

Lautenschlager, G. F., Lance, C. E., & Flaherty, V. L. (1989). Parallel analysis criteria: Revised equations for estimating the latent roots of random data correlation matrices. *Educational and Psychological Measurement, 49,* 339–345.

Levine, M. S. (1977). *Canonical analysis and factor comparison.* Beverly Hills, CA: Sage.

Lunneborg, C. E. (2000). *Data analysis by resampling: Concepts and applications.* Pacific Grove, CA: Duxbury.

MacCallum, R. C., Browne, M. W., & Sugawara, H. M. (1996). Power analysis and determination of sample size for covariance structural modeling. *Psychological Methods, 1,* 130–149.

MacCallum, R. C., Widaman, K. F., Zhang, S., & Hong, S. (1999). Sample size in factor analysis. *Psychological Methods, 4,* 84–99.

Marcoulides, G. A., & Schumacker, R. E. (2001). *New developments and techniques in structural equation modeling.* Mahwah, NJ: Erlbaum.

Meredith, W. (1964). Canonical correlations with fallible data. *Psychometrika, 29,* 55–65.

Montanelli, R. G., Jr., & Humphreys, L. G. (1976). Latent roots of random data correlation matrices with squared multiple correlations on the diagonal: A Monte Carlo study. *Psychometrika, 41,* 341–348.

Mulaik, S. A. (Ed.). (1992). Theme issue on principal components analysis. *Multivariate Behavioral Research, 27*(3).

Mulaik, S. A., James, L. R., Van Alstine, J., Bennett, N., Lind, S., & Stillwell, C. D. (1989). An evaluation of goodness of fit indices for structural equation models. *Psychological Bulletin, 105,* 430–445.

Nasser, F., Benson, J., & Wisenbaker, J. (2002). The performance of regression-based variations of the visual scree for determining the number of common factors. *Educational and Psychological Measurement, 62,* 397–419.

Nasser, F., & Wisenbaker, J. (2003). A Monte Carlo study investigating the impact of item parceling on measures of fit in confirmatory factor analysis. *Educational and Psychological Measurement, 63,* 729–757.

Nunnally, J. C. (1978). *Psychometric theory* (2nd ed.). New York: McGraw-Hill.

O'Connor, B. P. (2000). SPSS and SAS programs for determining the number of components using parallel analysis and Velicer's MAP test. *Behavior Research Methods, Instruments, and Computers, 32,* 396–402.

Ogasawara, H. (2000). Some relationships between factors and components. *Psychometrika, 65,* 167–185.

Pedhazur, E. J. (1982). *Multiple regression in behavioral research: Explanation and prediction* (2nd ed.). New York: Holt, Rinehart & Winston.

Russell, D. W. (2002). In search of underlying dimensions: The use (and abuse) of factor analysis in *Personality and Social Psychology Bulletin. Personality and Social Psychology Bulletin, 28,* 1629–1646.

Satorra, A., & Bentler, P. M. (1994). Corrections to test statistics and standard errors in covariance structural analysis. In A. von Eye & C. C. Clogg (Eds.), *Latent variable analysis: Applications for developmental research* (pp. 399–419). Thousand Oaks, CA: Sage.

Schmid, J., & Leiman, J. (1957). The development of hierarchical factor solutions. *Psychometrika, 22,* 53–61.

Schumacker, R. E., & Marcoulides, G. A. (1998). *Interaction and nonlinear effects in structural equation modeling.* Mahwah, NJ: Erlbaum.

Snook, S. C., & Gorsuch, R. L. (1989). Component analysis versus common factor analysis: A Monte Carlo study. *Psychological Bulletin, 106,* 148–154.

Spearman, C. (1904). "General intelligence," objectively determined and measured. *American Journal of Psychology, 15,* 201–293.

Steiger, J. H., & Lind, J. C. (1980, June). *Statistically based tests for the number of common factors.* Paper presented at the annual meeting of the Psychometric Society, Iowa City, IA.

Stephenson, W. (1953). *The study of behavior: Q-technique and its methodology.* Chicago: University of Chicago Press.

Tatsuoka, M. M. (1971). *Multivariate analysis: Techniques for educational and psychological research.* New York: Wiley.

Thompson, B. (1980a). Comparison of two strategies for collecting Q-sort data. *Psychological Reports, 47,* 547–551.

Thompson, B. (1980b). Validity of an evaluator typology. *Educational Evaluation and Policy Analysis, 2,* 59–65.

Thompson, B. (1984). *Canonical correlation analysis: Uses and interpretation.* Newbury Park, CA: Sage.

Thompson, B. (1988). Program FACSTRAP: A program that computes bootstrap estimates of factor structure. *Educational and Psychological Measurement, 48,* 681–686.

Thompson, B. (1989). Meta-analysis of factor structure studies: A case study example with Bem's androgyny measure. *Journal of Experimental Education, 57,* 187–197.

Thompson, B. (1990a). MULTINOR: A FORTRAN program that assists in evaluating multivariate normality. *Educational and Psychological Measurement, 50,* 845–848.

Thompson, B. (1990b). SECONDOR: A program that computes a second-order principal components analysis and various interpretation aids. *Educational and Psychological Measurement, 50,* 575–580.

Thompson, B. (1991). A primer on the logic and use of canonical correlation analysis. *Measurement and Evaluation in Counseling and Development, 24,* 80–95.

Thompson, B. (1992a). DISCSTRA: A computer program that computes bootstrap resampling estimates of descriptive discriminant analysis function and structure coefficients and group centroids. *Educational and Psychological Measurement, 52,* 905–911.

Thompson, B. (1992b). A partial test distribution for cosines among factors across samples. In B. Thompson (Ed.), *Advances in social science methodology* (Vol. 2, pp. 81–98). Greenwich, CT: JAI Press.

Thompson, B. (1993a). Calculation of standardized, noncentered factor scores: An alternative to conventional factor scores. *Perceptual and Motor Skills, 77,* 1128–1130.

Thompson, B. (1993b). The use of statistical significance tests in research: Bootstrap and other alternatives. *Journal of Experimental Education, 61,* 361–377.

Thompson, B. (1994). The pivotal role of replication in psychological research: Empirically evaluating the replicability of sample results. *Journal of Personality, 62,* 157–176.

Thompson, B. (1995). Exploring the replicability of a study's results: Bootstrap statistics for the multivariate case. *Educational and Psychological Measurement, 55,* 84–94.

Thompson, B. (1996). AERA editorial policies regarding statistical significance testing: Three suggested reforms. *Educational Researcher, 25*(2), 26–30.

Thompson, B. (1997). The importance of structure coefficients in structural equation modeling confirmatory factor analysis. *Educational and Psychological Measurement, 57,* 5–19.

Thompson, B. (2000a). Canonical correlation analysis. In L. Grimm & P. Yarnold (Eds.), *Reading and understanding more multivariate statistics* (pp. 285–316). Washington, DC: American Psychological Association.

Thompson, B. (2000b). Q-technique factor analysis: One variation on the two-mode factor analysis of variables. In L. Grimm & P. Yarnold (Eds.), *Reading and understanding more multivariate statistics* (pp. 207–226). Washington, DC: American Psychological Association.

Thompson, B. (2000c). Ten commandments of structural equation modeling. In L. Grimm & P. Yarnold (Eds.), *Reading and understanding more multivariate statistics* (pp. 261–284). Washington, DC: American Psychological Association.

Thompson, B. (2002). "Statistical," "practical," and "clinical": How many kinds of significance do counselors need to consider? *Journal of Counseling & Development, 80,* 64–71.

Thompson, B. (Ed.). (2003). *Score reliability: Contemporary thinking on reliability issues.* Newbury Park, CA: Sage.

Thompson, B., & Borrello, G. M. (1985). The importance of structure coefficients in regression research. *Educational and Psychological Measurement, 45,* 203–209.

Thompson, B., & Borrello, G. M. (1986). Second-order factor structure of the MBTI: A construct validity assessment. *Measurement and Evaluation in Counseling and Development, 18,* 148–153.

Thompson, B., & Borrello, G. M. (1987, January). *Comparisons of factors extracted from the correlation matrix versus the covariance matrix: An example using the Love Relationships Scale.* Paper presented at the annual meeting of the Southwest

Educational Research Association, Dallas. (ERIC Document Reproduction Service No. ED 280 862)

Thompson, B., Cook, C., & Heath, F. (2001). How many dimensions does it take to measure users' perceptions of libraries?: A "LibQUAL+™" study. *portal: Libraries and the Academy, 1,* 129–138.

Thompson, B., Cook, C., & Heath, F. (2003). Structure of perceptions of service quality in libraries: A LibQUAL+™ study. *Structural Equation Modeling, 10,* 456–464.

Thompson, B., Cook, C., & Thompson, R. L. (2002). Reliability and structure of LibQUAL+™ scores: Measuring perceived library service quality. *portal: Libraries and the Academy, 2,* 3–12.

Thompson, B., & Daniel, L. G. (1996). Factor analytic evidence for the construct validity of scores: An historical overview and some guidelines. *Educational and Psychological Measurement, 56,* 197–208.

Thompson, B., & Pitts, M. C. (1982). The use of factor adequacy coefficients. *Journal of Experimental Education, 50,* 101–104.

Thurstone, L. L. (1935). *The vectors of the mind.* Chicago: University of Chicago Press.

Thurstone, L. L. (1947). *Multiple factor analysis.* Chicago: University of Chicago Press.

Tucker, L. R. (1966). Some mathematical notes on three-mode factor analysis. *Psychometrika, 31,* 279–311.

West, S. G., Finch, J. F., & Curran, P. J. (1995). Structural equation models with nonnormal data. In R. H. Hoyle (Ed.), *Structural equation modeling* (pp. 56–75). Thousand Oaks, CA: Sage.

Wilcox, R. R. (1998). How many discoveries have been lost by ignoring modern statistical methods? *American Psychologist, 53,* 300–314.

Wilkinson, L., & Task Force on Statistical Inference, APA Board of Scientific Affairs. (1999). Statistical methods in psychology journals: Guidelines and explanations. *American Psychologist, 54,* 594–604. [Reprint available from http://www.apa.org/journals/amp/amp548594.html]

Vacha-Haase, T. (1998). Reliability generalization: Exploring variance in measurement error affecting score reliability across studies. *Educational and Psychological Measurement, 58,* 6–20.

Vacha-Haase, T., Kogan, L. R., & Thompson, B. (2000). Sample compositions and variabilities in published studies versus those in test manuals: Validity of score reliability inductions. *Educational and Psychological Measurement, 60,* 509–522.

Veldman, D. J. (1967). *Fortran programming for the behavioral sciences.* New York: Holt, Rinehart & Winston.

Zwick, W. R., & Velicer, W. F. (1986). Factors influencing five rules for determining the number of components to retain. *Psychological Bulletin, 99,* 432–442.

INDEX

ABOUT THE AUTHOR

Bruce Thompson is a professor and distinguished research scholar at the Department of Educational Psychology at Texas A&M University, adjunct professor of family and community medicine at Baylor College of Medicine (Houston), and professor of interdisciplinary health sciences at Louisiana State University Health Sciences Center. He is the editor of *Educational and Psychological Measurement* and the series *Advances in Social Science Methodology*, and past editor of two other journals. He is the author or editor of 9 books, 13 book chapters, 178 articles, 22 notes–editorials, and 11 book reviews. His contributions have been especially influential in moving the field with regard to greater emphasis on effect size reporting and interpretation, and promoting improved understanding of score reliability. He is a fellow of the American Psychological Association and an elected member of the Council of the American Educational Research Association.